Managing Writers

A Real World Guide to Managing Technical Documentation

Richard L. Hamilton

Managing Writers

Copyright © 2009 Richard L. Hamilton

Some sections of this book have previously appeared in the author's blog [http://rlhamilton.net/blog] and on The Content Wrangler web site [http://thecontentwrangler.com].

Disclaimer

Trademarks

XML Press
Fort Collins, Colorado 80528
http://xmlpress.net

First Edition
ISBN: 978-0-9822191-0-2

Printed in the United States of America.

To Mei-li

Table of Contents

Preface

When I started working as a documentation manager, common "wisdom" was that anyone could manage technical documentation. I had no experience with documentation or management, and no one, except possibly the people I was to manage, thought that would be a problem. As I soon discovered, managing documentation takes a broader set of skills than most engineering management jobs. Where else would you find yourself managing a team that included a carpenter, electrical engineer, English major, psychologist, and mathematician, all working together on software documentation? That is not a random assortment; that is the range of skills on just one of the documentation teams I have managed.

In addition to being able to manage a diverse team, a documentation manager must have a strong technical background, both to understand the technology being documented and to understand the technology behind the tools being used by the documentation team.

If that is not enough of a challenge, consider that technical documentation lives at the intersection of two disciplines, engineering and communicating, which are the oil and water of the business world. You and your writers need to bridge that gap internally, with engineers, and externally, with customers.

Most important, technical documentation is viewed with disdain by many engineers and lives at the bottom of the power hierarchy in most companies. A significant amount of your time as a documentation manager will be spent working to gain respect, power, and leverage so you can do your job.

These factors make managing technical documentation challenging and frustrating, but also rewarding. Documentation has broadened from a discipline focused on writing books to one that communicates through words, graphics, audio and video, in media from paper to help systems to web sites to podcasts and beyond. Despite the perception that "no one reads the documentation," the right documentation in the right form can make the difference between a product that works and one that does not.

This book covers the basics of managing a documentation team, including hiring, motivating, and planning, and it goes beyond the basics to look at the things that make this discipline unique. My objective is to give you the information you need to successfully manage documentation people, projects, and technology.

Audience

This book is for anyone, regardless of title, who manages technical documentation projects or people. In addition to those who hold the title "Documentation Manager," this includes:

- ► Writers who manage their own work or the work of others. In a down-sized world it is common for writers to be expected to manage themselves and their schedules. This book will help you plan your projects better and be more effective working with your engineering teams.

- ► Product development or marketing managers who have one or more writers on their team. Even if you delegate a lot of the management job to those writers, this book will help you understand their challenges and be supportive of their needs.

Structure

The book has four major parts: Getting Started, Managing People, Managing Projects, and Managing Technology.

Getting Started introduces the book. Chapter 2: *The Elements of Technical Writing* covers the essential elements of technical documentation from the writer's perspective. Chapter 3: *Power and Influence* discusses the central struggle for documentation managers, power.

Managing People discusses motivation, change management, performance evaluation, hiring, and human resources. Despite the existence of references in all of these areas, many managers start their careers woefully ignorant of how to manage people. In addition to looking at the basics of good people management, I also look at some of the things that make managing technical communicators a unique challenge.

Managing Projects discusses development methodologies, project planning, estimation, tracking, and localization. Managing technical documentation projects shares some characteristics with managing projects in other disciplines, but it also offers unique challenges. Most of these challenges result from the low position of documentation in the power structure. Documentation managers have less influence over schedules than managers in other disciplines and often find themselves trying to fit ten pounds of potatoes in a five pound sack. This section focuses on how to deal with these challenges effectively.

Managing Technology looks at the realities of technology for technical documentation. Technical documentation is often considered to be a *non*-technical discipline. In reality, not only do you need to deal with whatever technology is in your product or service, you need to deal with technologies that support your work. XML, Content Management Systems, and the Internet are all things you need to be concerned with, and you cannot expect much help from managers or engineers in other disciplines. The tools your team uses are distinct from those used in other disciplines. Understanding the possibilities and limitations is critical to your success. This section looks at how you can live with technology and acquire new technology. It also discusses technologies that support documentation, including XML, Content Management Systems, and the Internet.

Acknowledgements

Many people have helped make this book what it is. Thanks to everyone who read and commented on sections of this book through my blog [http://rlhamilton.net/blog], including: Jim Earley, Chip Gettinger, Judy Horton, Scott Hudson, Jim Leth, Mark Modig, Mike Ruscio, and Kate Shorey. Special thanks to Steve Bourgault, Jim Hamilton, Laura Praderio-Lynn, Larry Rowland, and Spence Wilcox for reading and commenting on large portions of this book.

Thanks to Garrett Long, who was largely responsible for convincing me to take my first job as a documentation manager. Thanks to Steve Bourgault, Bill Klinger, Sue Picus, and Ray Terry, who managed me during my time as a documentation manager; I learned a lot from each of you.

Thanks to John Hedtke for his wise counsel on writing and marketing, and to Scott Abel, the "Content Wrangler," who helped in many ways. Thanks to the DocBook community, especially Norm Walsh and Bob Stayton, for keeping the DocBook flame alive and providing many of the tools used to write this book.

Finally, thanks to my wife Mei-li, who supported me in every way possible while I wrote this book.

Getting Started

1

Introduction

> *There they go.*
> *I must run and catch up with them,*
> *because I am their leader!*
> — Mahatma Gandhi

After 10 years developing software, including a few years leading a team informally, I felt ready to become a project manager. I had been a successful software developer, writing system and user software, and had recently returned from an assignment in Japan where I initiated and led a large project for a major customer in China. In my not so humble opinion, I was prepared to lead a crack development team to greater glory.

The day my opportunity arrived, I was called in to meet the Lab Manager. This was back when managers actually played the part. He wore a Brooks Brothers dark grey pin-stripe suit with a red "power" tie. His window office had an immaculate mahogany desk with a beautiful Newton's cradle and a couple of other high end executive toys. The only papers on his desk were neatly lined up in mahogany In/Out trays. He managed several hundred engineers and was responsible for millions of dollars in projects.

He told me that he wanted to "explore the possibility" of offering me a job managing technical documentation for the UNIX operating system.

Managing technical documentation was far from my first choice, but the offer was not a complete surprise. The last two people to be promoted into management were first put in a documentation management job, then within a few months were given a lateral move to manage a software development team. It was commonly accepted that a promotion to documentation manager was a relatively safe way to let a manager get his or her feet wet; after all, how much trouble can you get into managing documentation?

This time, however, he wanted to stop the revolving door. He said he would only offer the job to someone who agreed not to seek another management position for at least two years. Had there been other openings, I never would have considered this job. My experience with documentation was sketchy at best. In fact, I barely recognized documentation as part of the process, let alone an essential part. But, there were no other openings, and it did not look like there would be any for quite a while, so I took the job.

That was nearly 20 years ago, and except for a couple of a short time outs, I have been managing technical documentation ever since. I have had the chance to do other things, but I keep drifting back to this field. I have found that managing technical documentation is both a challenge and great fun. Plus, I have had the opportunity to work with a group of interesting and talented individuals, who have taught me a lot.

By necessity, any individual's approach is constrained by his or her personal experience. While I have done my best to draw on the knowledge of others in the field, my approach reflects my skills and experience. My university training is in computer science and music, and I have continued to keep my technical skills up-to-date over the years, always having a software project or two and some music bubbling on the back burner.

My management experience is primarily in large companies, managing documentation for software products, some of them end-user products, some of them highly specialized system products, like the UNIX and Linux operating systems. In addition, I have managed several SGML/XML technology projects, including recent projects targeted at improved single-sourcing and content reuse using DocBook XML.

I believe the strongest management style is one that avoids tight control and encourages independence. I believe in light-weight processes, but thorough planning. I believe in treating writers like adult human beings, who nearly always know how to do their jobs better than I do. I believe in hiring the smartest and hardest working people I can find, and rewarding them for good work.

I believe the most important function of a manager is to set up an environment where writers can be productive; an environment where writers are respected, given the tools they need, shielded from interference from the corporation and its managers (including me), and left alone to do their jobs.

I believe in taking full advantage of technology, but having been seduced more than once by a hot new tool, I have developed a jaundiced view of what technology can and cannot do. As a member of the DocBook Technical Committee, I believe in the power of XML and DocBook, but I am also aware that XML is not the answer to every question about documentation technology.

Looking back on my first days as a documentation manager, I am amazed that anyone would entrust the technical documentation for a major product to someone so completely unprepared. Fortunately, the team I inherited was very experienced, used to clueless managers, and fully capable of getting along without a manager until they could train me.

And since I was enlisted for two years, they had some time to whip me into shape and were willing to spend a little more effort on a manager who was likely to be there longer than most. Even though I did not go into that job thinking it would be my life's work, I became a willing student, and I have never regretted taking that job. And, I am proud to still count the members of that first team among my best friends.

Over this time I have developed a set of strategies and tactics based on solid principles that have served me well. There is nothing trendy or exotic here; there are no silver bullets or easy answers. At the same time, there is no magic involved; if you approach the job with your eyes open and your ego in check, the chances are you will succeed.

2

The Elements of Technical Writing

Not just anyone, with any background,
or any training, can do a fine job of programming.
Programmers know this, but then why is it that they think
that anyone picked off the street can do documentation?
— Gerald Weinberg

This chapter discusses the elements of the technical writing job – those things practitioners deal with every day – and for each focuses on the one or two most important aspects. It is not a tutorial on technical writing; instead, it is a road-map of the things technical writers deal with every day and that their managers need to understand to effectively lead technical documentation projects.

The elements of technical writing are: product, developers, audience, tasks, deliverables, environment, and schedule. Along with strong writing skills, which are a pre-requisite, they comprise everything important that a technical writer needs to be concerned about. I will present them in roughly the order they need to be considered, but recognize that you and your writers need to deal with all of them throughout the course of a project.

The Product

The *product* is whatever you are writing about, even if it is not a product. It could be a service, software, hardware, an airplane, or a toaster. Whatever it is, your job is to understand the product. Read the specifications, technical requirements, marketing requirements, and documentation for related products. If you are updating documentation for an existing product, read the existing documentation.

Get your hands on the product. If you are documenting emergency procedures for the Space Shuttle, you may not be able to give it a test drive, but usually you can get your hands on a sample and try it out. Even with the Space Shuttle, maybe you can try the simulator or get familiar with it in other ways.

For example, the first technical documentation I ever worked on was a guide to using the Application Programming Interface (API) for a character-based user interface for the UNIX operating system

Before writing any documentation, I wrote some programs using the API. I just grabbed some examples written by others and compiled them; no *real* programming was required. But, I got a feel for what was going on, and I got a test-bed. That test-bed let me try out the API, find problems in the existing documentation, and test new development loads.

You will have to figure out the best way to learn your particular product, but however you do it, get as close as you can to the thing you are writing about.

The Developers

While it is critical to get close to the product, it is just as important to get close to the *developers* who are designing and building the product. Next to the product itself, they will be your most important source of information.

Your relationship with the development team, and especially any assigned contacts, is critical. If your contacts see you as adding value to the product, and even more importantly, if they see you as someone who understands and appreciates what they are doing, they will do everything they can to help you. If not, they will do everything they can to avoid you.

Here are a few of the things that I have found useful:

- ▸ Use the engineering team's time wisely:
  - Do not make them teach you anything you can learn from existing documents.
  - Prepare questions before you meet them.
  - Try to schedule reviews when they are not in a crunch – this is always difficult, but it is still worth the effort to try.
  - Do not ask for additional reviews unless they want them or there are significant changes from a previous draft.
- ▸ Be a visible part of the project. Attend planning meetings, participate in discussions about the project, have lunch with the team.
- ▸ Keep the project up-to-date on your progress.
- ▸ As a manager, keep in touch with your engineering management peers. Do not make your first meeting a cry for help.

If you build a *good* relationship with the engineering team, you will have a much easier time getting the information and cooperation you need to successfully complete your deliverables. If you build a *great* relationship, they will tell you about new features long before those features show up in a requirements document, they will warn you about problems that need documentation workarounds, and they will keep you up-to-date on the *real* status of the project.

The Audience

The *audience* is whoever you are writing the documentation for. Presumably, this is a group of people who will be using the product in some way. For some products identifying the audience is easy. With the API I mentioned above, the audience was programmers who used the API to create user interfaces for their software. For the Space Shuttle emergency procedures, however, the audience could include hundreds of flight crew and ground crew members in a wide range of specialties, each of whom may have a different role to play in an emergency.

Defining your audience is critical to defining the scope of your documentation, the deliverables required, and the resources needed. Therefore, you need at least a rough idea of the audience very early, even if that rough idea is nothing more than: "Homeowners who will use this product to shampoo carpets."

The project manager or marketing manager should know a fair amount about the audience, but they are unlikely to volunteer that information unless you ask the right questions. Here are a few you should get answered early:

- **Who will use the product?** What is the job role, or general activity of the person using the product? For example, system administrator, person shampooing carpets, or NASA communications specialist.

- **How will that person use the product?** For example, backing up data from a group of PCs, cleaning carpets, or communicating with the cargo mission specialist. At this point you just need a high level view; details can wait.

- **What is that person's background?** For example, entry-level to experienced system administrator, homeowner with no prior experience shampooing rugs, or senior electrical engineer with at least a Master's degree.

At some point in the process, though not necessarily at the very beginning, find time to talk with some of the audience. This is particularly useful if you are updating existing documentation and can talk with someone who already uses the product, but even for a new product, you will learn a lot if you spend time with likely users. If you can, give potential users draft copies of your documentation. This can be tough with some products, but where you can, you will get useful information. The better you know the likely users of the product, the better your documentation will be.

The Tasks

Tasks are the things your audience will do with your product. For example, setting up backup media, loading detergent into the carpet cleaner, or engaging the emergency communications system. These are the activities that your documentation will describe.

Most documentation these days is "task-oriented," which means you do a "task analysis" that identifies the tasks users will do with the product. Then, you organize the documentation around those tasks. Task orientation is a common and powerful way to organize your documentation. In most cases, users are not really interested in reading documentation; they want to accomplish a task. The quicker they can complete that task, the happier they will be.

However, be careful; task-orientation can be over-done. I have seen documentation that is simply a list of tasks with step-by-step instructions for each one. That is fine until you want to do something the writer has not thought about. Then you are stuck unless you have some broader context about the product and what it can do. Good documentation will provide that context and will provide other material to support the user.

The two major categories of supporting information are conceptual and reference. Conceptual information includes topics like: planning a backup strategy, understanding how your steam cleaner works, or knowing when to declare an emergency. Reference information includes things like: a list of command line options for the backup application, a chart for calculating how much shampoo to use for carpets of different sizes, or a table showing fuel tank capacities.

As a general rule, conceptual and reference material supports the task descriptions. Therefore, you usually define the tasks first, then determine what conceptual and reference information you need to support those tasks. Once you have the tasks and the supporting information identified, you can determine your deliverables.

The Deliverables

Deliverables are the recognizable things that writers deliver to the project. For example, User Guides, Administrator Guides, Help Systems, Quick Reference Cards, and Reference Manuals. Deliverables take many forms,including: printed books, glossy inserts, packaging copy, posters, web pages, wikis, and help systems. Deliverables are usually the smallest things tracked by managers outside your team, though they are not the smallest things you will need to track.

At one time, deliverables were the center of the writer's universe. They were defined in advance, and the writer, possibly with production and graphics help, created each deliverable as a unique item in one final form. While this still happens, you are just as likely to work in an environment where the writer's job is to create a set of more or less independent modules, often based on tasks. These modules are then assembled into deliverables late in the process, often by others. In a modular environment, any individual module may end up in several different deliverables, and any given deliverable may have modules from several different writers.

However, regardless of the process, you still need deliverables and someone still needs to design them. You need to select a set of deliverables that work for your product, select the content for each of those deliverables, and select an organization for that content. Modular methodologies can make that process easier and give you more flexibility than you might have had in an earlier era, but they cannot eliminate it.

How to design your deliverables is beyond the scope of this book. What matters here is that there are two different perspectives you need to consider regarding deliverables: internal, how your company sees the deliverables, and external, how your customers see the deliverables.

Your company sees deliverables as work items with schedules, deadlines, and often, individual budgets. Management will evaluate you on how well you meet your schedule, deadlines, and budgets. Often they will judge you more on these elements than on the content itself, though as you might guess, organizations where this is true usually produce substandard documentation.

On the other hand, customers only care whether your deliverables help or hinder them as they use your product. Their judgement depends on more than just "covering" the information. Even if all the information is there, badly chosen or badly designed deliverables make it much harder for customers to find what they need. You need to select the correct set of deliverables and then design each individual deliverable to be as effective as possible.

The balancing act between these two perspectives is to create a set of deliverables that will support your customers and will fit your company's internal needs. These internal needs include things like: existing documentation structure, expectations of the development team, and division of labor among writers. In my experience, this is not a difficult balancing act, but it does take some time and effort to do well.

The Environment

The *environment* is the set of tools, processes and support personnel that the writer works with. This could be as simple as Microsoft Word and a printer, or as complex as an ISO 9000 compliant process with a high-end Content Management System (CMS), staff to run the CMS, graphics designers, indexers, and production specialists. The environment also includes the skills and experience of each person working on the project.

Your environment, regardless of what it is, sets an upper limit on your team's productivity and shapes the types of deliverables you can efficiently create. If you understand the strengths and limitations of your current environment, you can select deliverables that you can efficiently develop and avoid ones that will cause heartburn.

Environments constantly evolve. Even if you stick with Microsoft Word, or some other standard word processor, you will need to upgrade your software and hardware periodically. Therefore, understanding your environment also means understanding how you want it to evolve. Where do you see the products you support going? What do those changes mean to your documentation? What impact will industry trends (new platforms, the Internet, etc.) have on your work? What opportunities do you see for improving your customers' experience with your documentation. What opportunities do you see for improving your team's effectiveness and efficiency?

You should always know where you want your environment to be in the next few months and years, and you should be planning how to get there. If you neglect this task, you will eventually find yourself with obsolete hardware and software, and a team unprepared to handle needed changes. You will invariably hit this wall when both the schedule and money are tight.

Evolution in the environment also includes developing your team professionally, for example by giving them opportunities for training. The skills of your team are more important than any tool you can buy. The bottom line is that you need to thoroughly understand the strengths and weaknesses of your environment, including your team, and be constantly improving it in manageable increments.

The Schedule

The *schedule* and deliverables are the most visible parts of your work inside your company. Therefore, while they may not be the most important aspect of what you need to deliver, they deserve careful attention.

Schedules are the closest thing to a "black art" that you are likely to deal with as a documentation manager. The good news is that as a documentation manager, you will rarely set schedules; the bad news is that you will rarely set schedules.

With the exception of the production back-end, documentation managers rarely have significant influence in setting project-visible schedules. Project managers understand that some time is required to produce documentation, paper or electronic, so even though they will push to make that time as short as possible, they will concede that *some* time is needed. However, I have never seen a schedule lengthened solely to give writers time to create better documentation, and I have never seen a project delayed by documentation, except when the back-end processes hit a glitch.

Therefore, documentation managers usually find themselves fitting their work into an existing schedule, rather than creating a schedule that fits their work. Chapter 10, *Project Planning* (p. 101) discusses scheduling in detail, including strategies for dealing with unreasonable schedules.

Putting it Together

Being successful as a technical writer or manager of technical writers means not only mastering these elements individually, but also mastering the interactions between these elements. It also requires proficiency in writing, planning, negotiating, listening, and communicating verbally.

Every situation provides a unique combination of elements and requires a unique response. While I try to give good general advice throughout this book, it is important to recognize that one size does *not* fit all. You will need to gain experience and apply that experience, making lots of mistakes, to gain proficiency.

If you are a new manager, especially one who has little or no previous experience in documentation, you might find it useful to think of these elements as checklists for asking questions and learning more about what your team is doing. You rarely can go wrong asking your writers to tell you about the audience, or what the product does, or how they plan to organize their deliverables, or whether they are comfortable with the schedule. In fact, if a writer or for that matter an interviewee cannot answer basic questions about these elements in his or her work, consider it a red flag.

3

Power and Influence

Never stand begging for that which you have the power to earn.
— Miguel de Cervantes

"You won't catch me delaying a product release because of the documentation," he said, in response to what I thought was a reasonable assertion that user documentation is part of the product and that the project schedule should reflect this reality. We were in a project meeting looking, as always, for some way to avoid yet another delay. And, as always, the question was whether we could shave some time off the back-end of the documentation schedule to compensate for adding time to the development schedule.

While you may not get quite as direct a rebuke from your project manager as I did, you will have to look hard to find one who would choose to delay a release just to improve the documentation. And, while I am sure it must have happened somewhere, to some now unemployed manager, I have never seen a product release delayed because of the documentation.

This does not mean you do not have the power to delay a product, but using that power is as close as you or I are likely to get to using the "nuclear option." You are going to need to throw yourself in front of the release train and convince the product manager that it is risking his or her career, not to mention yours, to put the product out without the right documentation.

Every documentation team I have managed or observed has been constantly under pressure to shorten schedules and reduce effort. Your ability to resist unreasonable pressure correlates directly to your ability to influence the project management team and your peer managers.

Your ability to influence others—your authority within the organization—derives from formal and informal sources. [1] Formal authority derives from your position in a management chain or your assigned role on a project. It is delegated to you – and can be taken away – by a higher formal authority, the company you work for and your management chain. The good news is that formal authority is well understood and accepted in the corporate world, and if you have the full weight of your management chain behind you, it can be a powerful force.

The bad news comes in three parts: first, formal authority does not help very much with those above you in the management chain, second, it can be taken away just as easily as it is given, and third, documentation managers do not have much formal authority anyway. One of the hard facts of the documentation life is that documentation managers live at the bottom of the management food chain with formal authority over little more than their direct reports.

While you need it, formal authority is only one part of the picture. To succeed, you need to develop informal authority, which is any authority that is independent of formal authority. This section uses a true story of a situation where one of my teams was able to increase its informal authority creatively (to protect innocent and guilty alike, no names are used, but this happened as described).

This was a large system software project that issued an update release every year or so. The project integrated modules from dozens of subprojects, each of which was responsible for contributing to a "Delta Document," which was maintained by the support organization. The Delta Document described in detail every planned change in the next release.

---

[1] John Kotter's *Power and Influence* [25] is the definitive text on this subject.

The Delta Document was critical for the support organization, which used it for planning, but it was a pain to develop, and because it was of little interest to the development team, it was poorly written from the start and never updated. Further, because each section was written in a word processing format with a loosely defined structure, it was difficult to combine into a single document, and once combined, the document was nearly impossible to use.

At the same time, our documentation team was developing "Release Notes," which described the product to the end user. Since one of our prime sources of input was that same badly written, out-of-date, poorly formatted Delta Document, we were often forced to go back to the development team to gather information we should have already had.

With sufficient formal authority, we could have mandated a process that would ensure a well written, up-to-date Delta Document. But, like most documentation groups, our formal authority was severely limited. We might also have escalated the problem to someone higher in the management chain. But, while effective, escalation falls into the category of begging.

Though I have been reduced to begging on more than a few occasions, I will, to paraphrase Cervantes, only beg when there is no other option. Fortunately, we found a way to solve our problem and increase our influence with the development organization without getting on our knees.

The first thing we recognized was that there was not just one problem. The documentation team had a problem, which I described above, but the development teams did not care about that problem. Their problem was that they were required to contribute to a Delta Document whose value they did not really understand, then they were required to give the same information to the documentation team for the Release Notes, which again was of little value to them. They appreciated the need to document their changes, but not the need to do it twice, in different formats, for different teams. Since their problem and our problem were not the same, they had little incentive to solve our problem, especially if the solution left their problem unsolved.

Once we understood the development team's concerns, we were able to find a solution that addressed both problems. We took responsibility for both the Delta Document and the Release Notes, and turned them into one document. We developed a web site that let developers bring up their section of the Delta Document for the previous release, edit it to be appropriate for the new release, and save the new version. The documentation team took that input and first prepared the Delta Document for the support team. Then, as the project progressed, we edited the content for updates, consistency, and style, then brought it back to the developers for review at regular intervals. For those reviews, the development teams just needed to look at the current state of their sections and flag changes.

We made the combined document available on the web throughout the project in a consistent and easily searchable form. The project management team got an up-to-date view of the contents of the release, the development teams got a streamlined input process, and the documentation team got well-formatted, up-to-date information from a development team that was much happier.

We did not get this for free. We needed to develop the web input page and deal with a couple of extra review cycles. Plus, we took responsibility for a document we had not previously owned, the Delta Document. But, the cost was mostly short-term, and paid for itself by reducing the effort it took us to extract information from the development teams; both the mechanical effort of converting information from badly formatted sources, and the human effort of dealing with unhappy developers who could not understand why we wanted the same information they had just given another team.

However, the most important gain was in status. By giving the project managers a more usable, up-to-date Delta Document, and at the same time improving productivity for both ourselves and the software developers, we showed we could add value to the overall development process beyond that provided by our documentation.

The Japanese use the term "giri" to refer to a moral obligation or debt. While "giri" is a subtle and complex term, an explanation of which is well beyond the scope of this book, the idea of favors given and received is basic to any culture. Giri takes this a step further because favors are not given in anticipation of a specific return favor. In fact, while it is understood that reciprocation is required, that is never spoken. Instead, a favor is offered as a gift signifying loyalty and respect for the other person's status or position.

By solving this problem, we showed our respect for the development teams in a way that did not demand a favor in return, but still incurred a moral obligation that must be honored at some point. We also raised our status and informal authority by showing that we could design and implement a useful web tool.

Power – both the lack of it in most documentation teams and strategies for increasing it – is a central theme in this book. Without power within your organization, you will not be able to deliver quality documentation. But, as I hope this example illustrates, you can acquire power well beyond what you are given formally.

Managing
People

4

Working with Human Resources

Few great men could pass Personnel.
— Paul Goodman

The "Good Old Days"

I have a long history with Human Resources (HR), or as it was once called, "Personnel." My father was a personnel manager, and he met my mother when they both were working in a personnel department, so I suppose you could argue that I owe my existence at least in part to personnel management. My mother went on to other work, but my father remained a personnel manager throughout his career, which began in the forties and continued until the early seventies.

When my father spoke about his job, he spoke about employees. He worked with both white and blue collar workers and spent a lot of time in factories and offices talking with people and helping them with both work-related and personal problems. He knew everyone who worked in the divisions he supported and often knew their families, too.

My father was always interested in new methods and techniques and would occasionally bring home tests, usually psychological tests that were used for employee screening, and try them out on my brother and me. The one thing he never talked about was paperwork, and he never mentioned legal issues. I am sure both were part of his job, but they certainly were not the main focus.

Now, before we get lost in a Norman Rockwell painting, it is important to remember that when he first started working in the forties most of the workers were men, women were not paid equally, discrimination was legal, tests were unconstrained, and quotas were common for all sorts of things, from ethnic background, to religion, to what college you went to. But, it was also an environment where he could spend time working with people, rather than laws. I know that is why he chose that career, and I think that is still one of the reasons people go into HR.

The HR World Today

A lot has changed since then. Thankfully, in the United States, most forms of employment discrimination are illegal, pay for men and women is closer to equality, and pre-employment tests must relate to job performance. Our country and our workforce have become more diverse and working conditions are much fairer. These are not gains that we would want to give back, but they have not come without a cost.

The price for these gains has been tighter regulation of hiring and employment practices, increased scrutiny by the government, increased paperwork, and possibly most damaging, a greater distance between employees and management. That distance carries over to HR. Many employees only see an HR person on their first and last days of work, and those encounters are usually tightly choreographed exercises in filling out paperwork.

To an alarming degree, HR people spend their days navigating a labyrinth of legal landmines. For example, years ago I applied for a job at a large computer company. By complete chance, the HR person I spoke with was a woman I knew in college, but had not seen in years. We had plenty to catch up on, but because of the constraints of the interview process, she could not ask me anything that was not job related. So, the "how are you, what have you been up to for the last 10 years," conversation you would normally have with an old acquaintance was off-limits, and I am pretty sure that the information I volunteered so that she would not have to ask probably still made her nervous. That is the unfortunate reality of being an HR person these days. Most of the laws are there for good reason, but that does not make the job any easier.

Corporations are so concerned about exposure to law suits that they sometimes do crazy things. At one company I am familiar with, which was in the midst of layoffs, an employee who was nearing retirement and who was planning to leave the company soon anyway, volunteered to be laid off so that some other employee would be passed over. When the list of people to be laid off was revealed to managers, that employee was left off the list, but a colleague who was much younger was put on the list. The HR representative said that they needed to keep a particular distribution of ages in the group of people being laid off to avert the possibility of an age discrimination law suit. Having a skewed distribution might lead to a fairer outcome for the people involved, but could open up a crack for someone else to sue the company. Eventually things were straightened out, but it took a fair amount of wrangling, and it could easily have gone the other way.

The point here is not to disparage discrimination laws or the HR organization. I am a strong supporter of both. I was glad that the HR representative I met in the first story stayed within the letter of the law, even with an old friend. And, even though the second situation skirted insanity, in a perverse way I appreciate the diligence shown by that company.

However, the bottom line is that HR departments spend much more time than ever before dealing with legalities. Severe downsizing compounds the problem, leaving most HR departments with little time to interact with the humans they support. As a result, the days when an HR manager had the luxury of getting to know all of the employees he or she supported are long gone.

That said, you still should cultivate your HR people. In the second story, the manager's good working relationship with HR was critical. If they had not developed trust and mutual respect, the result would almost surely have been different.

When you *Must* Work with HR

The obvious, but still most important, reason for working with HR is that those laws that I have been talking about are targeted at you, and the HR folks are there to keep you, and the corporation, out of legal trouble. While HR can be helpful for any personnel issue, they will be essential when it comes to the hardest decisions: hiring and firing.

For hiring, they can help you make sure your screening and interviewing processes are fair and legal. Hiring is not just a matter of grabbing a few good looking résumés, asking a few random questions, then picking the person who makes the strongest first impression. In fact, that process pretty much guarantees you will make the wrong choice. If you screen and interview correctly, you will not only avoid legal problems, you will have a much better chance of hiring the right person. See Chapter 5, *Hiring* (p. 31) for a detailed discussion.

When it comes to firing, HR is even more important. We are decades past the point where Mr. Dithers could fire Dagwood on a whim. Most employment is "at will," meaning either the employee or the employer can terminate employment at any time, for pretty much any reason.

However, in practice, terminating employment for anything other than a gross violation of company policy or law almost always involves a script that must be followed to the letter. That script will normally involve documenting problems, working with the employee, developing and tracking improvement plans, delivering a sequence of warnings, and then only if all of these steps fail, terminating employment. Make the wrong step anywhere along the way and there will be legal problems. If you find yourself with an employee who is headed down this road, you will probably spend a lot of time with your HR representative.

Maintaining a Good Relationship with HR

Even though you are most likely to need HR during a crisis, you should spend time with your HR representative regularly. I always make it a point to meet my HR representative as soon as I take a new position or a new person gets assigned. And, I periodically find a reason to call or exchange email. For a long time, my HR representative was at a remote location that I regularly visited, and I made it a point to drop in whenever I was in town. I was surprised when she told me that many managers never call or drop in until there is a crisis.

This is not a good idea for a lot of reasons. First of all, if I have to deal with a difficult situation, I would prefer to speak with someone I know rather than a stranger. That is just a question of comfort level, but it is much easier to have a useful discussion when you have that level of comfort.

Secondly, someone you know will tend to give you the benefit of the doubt; he or she will want to be on your side, even if you have made a mistake. Of course, this only works if you have given that person good reasons for trusting you.

Thirdly, you will have a much better idea of how much you can reveal safely. Nearly every HR person I have worked with has taken confidentiality very seriously. But, you may run into that one person who is willing to reveal confidential information. Or, more likely, you may run into someone who has the tendency to reveal things inadvertently. You should never treat an HR person like a priest; even with the best of intentions he or she may reveal something you do not want revealed. But, if you have spent time with your HR representative, you will have a pretty good idea about what you can safely say.

Finally, HR has its finger on the pulse of the organization. You can learn a lot about what is going on just by keeping in touch with your HR representative. I do not mean you will pick up confidential information, which in my experience you probably do not want anyway. What you can get is information from a fresh perspective.

For example, if you are about to start a project with a manager you have not worked with before, most HR representatives will be happy to answer questions like, "What can you tell me about this person's management style?" You will not hear about the pending harassment suits or the reprimands on his or her permanent record, but you will get useful, if sanitized, information. And if you have a good enough rapport, the chances are that you will get a good sense of what he or she thinks about that person.

I know managers who have no time for HR. They may have read too much Dilbert and take Catbert, the "evil director" of human resources too seriously. Or they may see HR as a questionably necessary, time-wasting evil, and HR people as clueless drones who have no idea what is really going on.

If you feel that way, please think again. HR is there to help and generally does not have time to waste your time. They are also generally very personable folks, who love to chat when they can find the time. After all, you generally do not go into a Human Resources job if you dislike humans. But, even if you do not want to make your HR representative your best pal, take the time to meet him or her, get a feeling for what he or she is like, and stay in touch. I can pretty much guarantee that some day you will be glad you did.

5

Hiring

I don't hire anybody who's not brighter than I am.
If they're not brighter than I am, I don't need them.
— Paul "Bear" Bryant

For most managers, hiring causes nearly as much dread as firing. It is not the process itself, after all, most managers like to meet new people and talk with them. The problem is that among the decisions managers need to make, hiring is the one that can lead to the nastiest, longest lasting, and most painful mistakes. Of course, a good hiring decision can make a huge positive impact on your organization, but it is the bad decisions that haunt most managers.

Hiring technical writers can be especially challenging. Technical writers frequently come from unusual academic backgrounds that tell you next to nothing about their level of skill. Many have had careers that bounced all over the map; I have hired former carpenters, musicians, mathematicians, and sociologists. In fact, over ten years of hiring technical writers, I have hired only a handful who had a degree in technical communication.

This chapter looks at the qualities that make a great technical writer and how to evaluate those qualities when hiring. It also looks at the broader questions of staffing including: how to evaluate your staffing needs and how to balance your staff among employees, contractors, and service providers.

What Makes a Great Technical Writer?

The qualities that make a great technical writer mostly boil down to the ability to understand and the ability to communicate. A writer first needs to understand whatever he or she is writing about. That does not mean that technical writers need to be able to write computer software or design a widget, but it does mean that they need to understand how that software or widget is supposed to work for the user. No amount of skill with FrameMaker or Microsoft Word can compensate for an inability to understand the subject being described. I have never had to fire a writer for lack of tools skills, but I have had to for an inability to understand the subject matter.

A great technical writer will understand a product or service from the user's point of view. Often, product developers do not fully understand how people will use their products. They are too close to the product and understand the internal workings too well to have the same perspective as the people who actually buy and use the product. Good technical writers have the detachment needed to see a product or service from the user's perspective and to understand what the user needs to take full advantage of it.

In addition to understanding the product or service they are writing about, technical writers need to be able to communicate. This breaks down into two broad categories, communicating with the engineering team or product developers, usually verbally, and communicating with the end user, usually through writing.

We normally think first about writing ability, but the ability to communicate verbally with engineers is equally important. Since engineers are rarely rewarded or punished for their ability to work with the documentation team, this can be a real challenge. A technical writer who cannot extract information from the engineering team will fail, regardless of his or her writing skills.

Technical writers should be able to discuss a product or service with the developer and extract the information that a user will need to know. They should be able to do this without taking too much time, which to most developers is nearly any amount of time, and without going over the same information more than once. They should also be able to ask the right questions to get to the core of what they need with the least amount of effort.

Writing ability is of course central. The best technical writing empowers users and makes them feel smart. It also gives users a good impression about your product or service and your company. This is not easy to do, and it has become more difficult as writing teams adopt modular and single-source methodologies that require writers to structure information so that it can be combined in multiple ways and delivered in multiple media.

The best technical writers have the same knack that good journalists have; they get to the point and deliver what the reader needs to know concisely and directly. A good technical writer is not necessarily a skilled prose stylist; in fact, the wrong kind of style can be distracting. Instead, his or her writing is characterised by its directness, clarity, and conciseness. In addition, larger structures – books or large websites – will be clear and easy to navigate.

Overall, the best technical writers I have worked with are direct, but polite people. They are quick to understand a product or service, work well with busy, distracted engineers, and write clearly and concisely. They always have good work habits, are assertive, and have a sense of humor. The latter may not be essential, but it makes everything go a little more smoothly.

Evaluating a Candidate

Evaluating a technical writing candidate is similar in many ways to evaluating any other candidate. You need to evaluate the résumé, have an interview, and check references. With technical writers, you also need to evaluate writing samples. While the external process is largely the same, there are things you should be doing at each step that are unique for technical writers.

Evaluating a résumé

Most hiring managers, and I am no exception, take a couple of passes when reviewing résumés.   The first pass eliminates people who are clearly not what I am looking for. I try to answer two questions: first, are this person's qualifications even in the ballpark for the job, and second, can he or she write at least well enough to create a competent résumé?

Answering the first of these questions is not hard, but it does require that you understand the job requirements thoroughly and read the résumé closely. The former should be easy, but I have seen cases where a manager hired someone because he or she was a whiz at task A, when skill at task B was what was really required. I suggest that you document the job requirements in writing.

Read the résumé closely. A few years ago I needed to hire a tools person. A contract house I had worked with before said they had just the person I needed. I did not have much time, so I just glanced at the résumé before the interview. About five minutes in, the candidate stopped me and said, "Why are you interviewing me for this job, I'm not a tools expert"?

The contract house had obviously skimmed her résumé, seen the names of a few tools, and assumed, either honestly or possibly shaving the facts, that I might hire her as a tools person. Fortunately for her, I was also looking for writers, and I was so impressed with her honesty and outspoken manner that I kept interviewing her and ultimately hired her as a writer.

While that situation turned out well for everyone, I have found over the years that it is not at all uncommon to get a résumé with a job objective of "Technical Writer" when it is clear the person has no writing experience. In the worst cases, it can look like the objective was just slapped onto an otherwise unrelated résumé. Sometimes it is wishful thinking and sometimes it is a misunderstanding of what skills and experience are needed to be a technical writer, but either way it raises a red flag.

Whenever I see a résumé like that, it goes into the discard pile unless there is some indication that the person *can* write, and just as important, *wants* to write. While I have hired engineers with no prior technical writing experience, it was only after I became convinced that they were good writers who had solid reasons for a career change, and just as important, that they were not looking for an "easy" job after failing as an engineer.

The second thing I look at in that first pass is the way the résumé reads; after all, it is the first writing sample you will see from this person. If a candidate cannot communicate clearly and concisely in a résumé, there is no point in going further. Some managers will reject a résumé that contains even a single typo. I do not go quite that far, but I will reject a résumé that is badly written or contains blatant mistakes that a good copy-edit should catch.

I do not worry too much about the visual look of a résumé unless I am hiring for a job that requires design skills. At the same time, an unattractive visual design is at least a warning flag. There are plenty of attractive styles available in the standard word processors, so any résumé should be legible and look professional. I also do not care too much about whether the résumé is in chronological order or some other order, as long as all the information is there.

If the résumé is not eliminated by this point, I go back and look for the following things:

- **Writing experience:** It is surprisingly common to get résumés for technical writers who have no experience writing documentation. This is sometimes obvious, as for people who are moving from some other career into technical writing, but sometimes it can be covered up. Unless you are looking for an entry level writer, make sure you see real technical writing experience.

- **Subject matter familiarity:** The type of experience a writer has is at least as important as the amount of experience. Generally, I am looking for experience that shows an ability to understand similar concepts with a similar level of technical complexity. I do not expect a perfect match; I will consider someone who only has experience with Windows software for a project based on Linux software. However, I would be reluctant to hire someone whose sole experience has been with software to write the hardware maintenance manual for an aircraft, unless there are other indications – maybe a pilot's licence – that point to subject matter expertise.

- **Deliverable type experience:** The types of deliverables – books, web pages, online help, and so forth – a writer has worked with can also be a useful indicator, especially if you are working with one of the newer technologies, like XML. But most experienced writers have worked with several different kinds of deliverables, so I consider this less important than subject matter familiarity.

- **Tools experience:** Tools familiarity is another thing I look for, though I place less importance on it than some. In particular, unless I need someone with specific tools knowledge – for example, a FrameMaker guru – I am less concerned with the specific tools and more concerned with the type of content the tools work with. For example, if I am hiring someone for a team that is writing XML, I will give extra points to someone who has XML experience over someone who only has WYSIWYG experience, even if the person with XML experience has not used the same tools that my team uses.

One thing to be cautious about, and to explore when you get to an interview, is the level of skill people have with particular tools. Candidates know that recruiters will mechanically search résumés for the names of the tools their clients use, and if a résumé does not contain a reference to a desired tool it may be rejected. So, candidates will list tools they barely know, or that they hope they can learn in the weekend before a job interview. My rule of thumb is that if someone has not done a real project with the tools you are using or a tool in the same category, there will be a learning curve.

- **Education:** I do not worry too much about what degree a writer has, but I like to see a college degree of some kind. If not, there needs to be some strong experience in lieu of a degree. I am less concerned about where the degree was obtained or the GPA, though a degree from a diploma mill or a poor GPA will raise a red flag.

- **Leadership:** I like to see examples of leadership. This does not need to be formal or work related, it might be leading a small project, managing a budget, leading a Girl Scout troop, or organizing a fund drive. Even if you are not looking for a manager, I prefer people who can take the initiative and lead when necessary.

- **Other things to look for:** If the résumé has a section like "Other Qualifications," "Community Activities," or something similar, look at it closely. Skills and activities are included on a résumé because the candidate thinks they are important. Therefore, I always ask about them during the interview.

I also look for overlaps or gaps in experience, unusually short periods of employment, or unusual claims – for example, a technical writer who claims to have brought in $10 million in new business. That does not mean that any of these things are automatically disqualifying, but you need to explore them in the interview and the reference check.

Checking the web

Now is a good time to drop the candidate's name into your favorite web search engine. I suggest starting with Google, Zoominfo, and LinkedIn. Everyone knows Google, but you may not know that in addition to broad web searches, Google offers more targeted searches, including blogs, groups, news, patents, and books.

Zoominfo [http://zoominfo.com] crawls the web looking for information about people and assembles profiles based on that information. It then encourages people to "claim" their profile and add information to it. In my experience, it is worth searching Zoominfo, but if a person has not claimed his or her profile or has a common name, the results will be spotty.

LinkedIn [http://linkedin.com] is an online network of professionals. Unlike Zoominfo, which gathers information independently, all information in a LinkedIn profile is supplied by that person. That said, unless the person spends a lot of time on his or her profile, you will probably find gaps and missing information. That does not concern me unless I see blatant inconsistencies.

While they can be useful, there are a lot of caveats to consider when doing a web search. First of all, make absolutely sure that you are getting information for the person you are interviewing and not someone with a similar name. This is going to be difficult if your candidate is John Smith or Mary Johnson, but it can be tough with anyone. For example, if you drop my name into Google, you might decide that I am a basketball player with the Detroit Pistons, a fictional character from early 20th century adventure stories, a British artist, or a big-band drummer. And that does not include the dozens of Richard Hamiltons in the computer business.

To get the right person, you may need to narrow your search. You can do that by adding key words to your search; for example, adding "technical documentation" to a search for my name brings up my blog as the first hit, ahead of the basketball player, the artist, or any other namesake. Or, you can limit the search to websites or newsgroups that are devoted to relevant topics.

Narrowing your search will also help you avoid information that should be irrelevant (see Legalities (p. 39)). While it may be entertaining to read about a candidate's participation in the local Gay pride parade or his or her opinions on the president, I consider that kind of information to be irrelevant to job performance and using it to discriminate is unethical if not illegal.

The kinds of things I am looking for in a web search are things that confirm or contradict the résumé or that give me additional relevant input. For example, if she lists a publication, you might find a copy on the web. Or, you can verify whether he really gave the keynote speech at the 1993 STC Conference. You can also get a sense of someone's interaction style if you look at newsgroup postings or wiki entries. All of these things can, if used properly and legally, give you a more fully developed picture of a prospective employee.

Conducting a phone screening

Once I have decided that someone is probably a good candidate, I like to have a phone interview. Since a thorough in-person interview can take a day or more, a pre-screening by phone can help you narrow the field to a manageable size. It may also cause a candidate to withdraw early, again saving you time. And, it can help you prepare for the in-person interview by giving you clues as to where you should probe further and how you can approach the selling part of the interview.

Prepare for the interview the same way you would prepare for an in-person interview. Review the candidate's résumé before you call and have it in front of you during the call. Make a list of questions; I often write the questions on the résumé or circle the things I want to probe.

Make your questions as open-ended as possible; for example, instead of asking, "did you like your last company?," ask, "what did you like about your last company and what would you have changed?" I like to ask broad questions like, "tell me about the XYZ project you worked on at ABC company," then use the response to decide where to dig deeper.

Let the candidate talk and give him or her the chance to ask questions. You will not learn much listening to yourself talk. The questions a candidate asks can tell you a lot. For example, you should be able to tell how carefully he or she has researched your company and you; a sharp candidate will check out both. And, you should be able to judge how interested he or she is in the job.

Take notes as you go along. It will help you avoid asking the same questions again in the in-person interview, and it will serve as a memory aid when you go back over the ten phone interviews you did last week. It is easy to forget who it was you wanted to question further on a particular topic when you do not have notes and those ten interviews have faded into a blur.

Your goal is to decide whether to interview this candidate in person, so once you have made a decision, you can begin to wrap things up. If you are going to offer an interview, schedule it now if you can, if not, say so, and if you are not sure, say you will call back. Whatever you do, do not leave a candidate waiting for a response any longer than necessary; it is not fair to him or her and it makes you look bad.

Legalities

Before you communicate with a job candidate, make sure you understand the legal issues involved with hiring.

In the United States it is illegal for you to ask questions about age, race, birth place, gender, marital status, and disabilities, among other things. In other countries laws vary; make sure you know the laws that apply to you.

If a question cannot be related to job responsibilities, it may be illegal.

Conducting an interview

If you have done a good job evaluating résumés and screening by phone, you should have a group of qualified candidates. The job at this point is to choose the best among them.

Selecting the interview team

You should have several people interview a prospective employee. A weak candidate may get lucky and impress a single interviewer. But, it is much harder to impress multiple interviewers. Therefore, I want several positive interviews, and preferably 100% agreement, before hiring someone.

Interviews can range from a casual hour with the manager to several intense days with everyone who might have any contact with the candidate. In practice, somewhere in between should be adequate. I like to schedule a full day, including one-on-one interviews with as many members of my team as possible.

I like to have the same group of people do all of the interviews for a particular position. That is the best way to get a good comparison. I also try to get the people who will be working most closely with the candidate on the team. If you can get a member of the relevant development team as an interviewer, that is even better, though in many organizations that will be impossible.

Pick interviewers who will reflect positively on your organization; selling is as important as evaluating. I once interviewed with a company that had me talk with someone who was seriously unhappy with the company and its management, and he let it show. I learned stuff about that company they could not possibly have wanted me to hear. As the interviewee, I was glad to get his unvarnished opinion, but as an interviewing manager, I would never put someone with such a negative attitude on an interview team.

On the flip side, I chose my first job in part because one of the interviewers was enthusiastic about his job, and I wanted to work with him. We worked together for several years, and he is still a good friend. Having your top people on an interview team sells your company better than any brochure or glossy ad.

Preparing the interview team

Before starting a series of interviews, I like to get the team together to make sure everyone understands the requirements for the position being filled, understands his or her interview role, and has copies of résumés, writing samples, and other supporting material. I also want to make sure everyone understands the ground rules for conducting a fair and legal interview.

I like to assign roles to the members of an interview team. For example, I will ask one person to dig into the candidate's understanding of subject matter, and another to dig into tools knowledge. I may ask someone else to concentrate on probing a candidate's ability to solve problems. As long as you stick to skills and background relevant to the job, it is fine to pose hypothetical situations and ask candidates to solve problems. Just make sure you use the same process for each candidate and select situations and problems that are clearly related to the job requirements. This will help ensure fairness and give you a basis for comparison.

Two other important roles are the "host" and the "seller." The host is the person who meets the candidate at the beginning of the day, gives him or her the agenda, and acts as a point of contact. The seller is someone who sells the company and the job to the candidate. Often these roles are held by the same person, and I often take them on myself as manager.

There is nothing to stop the seller from evaluating the candidate or an evaluator from selling the company, but giving each interviewer a distinct role helps to ensure a uniform process without gaps. It also helps to make sure that interviewers cover the same topics with each candidate so they can make informed comparisons.

Evaluating candidates

Your job as you evaluate candidates is to find people who have the "great technical writer" skills described earlier, who can apply those skills to your needs, and who want to work in your organization. To that end, I like to balance the interview between an examination of the candidate's past accomplishments, an evaluation of his or her skills, and a sales pitch for the company and your team.

I usually follow the résumé in examining past projects and accomplishments. I go into more depth than the phone interview, but still ask open-ended questions related to the points outlined in the résumé. I will often select one or two projects as a vehicle to probe into a candidate's experience. Here are some questions you can use when questioning a candidate about a project:

- ▶ What was the candidate's role in the project? Résumés are often vague on the candidate's role. Was he or she working alone, as a leader on a team, or as a bit player?

- ▶ What skills did the candidate use? This includes subject matter skills, writing skills, and tools skills.

- ▶ What were the deliverables and what methodologies and tools were used to create those deliverables?

- ▶ What were the interactions with the development team like? Were there any particular problems the candidate had to solve?

- ▶ What was the result of the project? For example, was it delivered on time and did it satisfy customer needs? If not, why not? And, regardless of the result, what did the candidate learn from this project?

- ▶ Is this project represented in the candidate's writing samples? If so, you may want to look at the sample and ask some questions based on what you see. If not, ask why this project is not represented by a sample.

I may not use all of these questions; usually, I let the responses tell me where to go. So, if the project was a debacle, I focus on what the candidate learned and what he or she is doing differently as a result. If the project offered new challenges, like a new deliverable format or different subject matter, I will focus on how the candidate faced those challenges. Overall, what I want to see is how the candidate approaches projects, how deep his or her skills are, and what kind of attitude he or she brings to challenges.

You can get a sense of a candidate's skills through a discussion of his or her accomplishments, but it is a good idea to dig deeper. One useful way to dig deeper is to ask the candidate to perform a technical writing task during the interview. For example, you might give the writer a problem to solve, like, "write a procedure for re-filling the office coffee machine," and have the interviewer act as a subject matter expert. An exercise like this helps you evaluate the candidate's skills working with subject matter experts, as well as his or her organization and writing skills.

Once you have completed a set of interviews with a candidate, make sure you collect feedback quickly; after even a day, memory fades. As part of the feedback, I want to know if anyone has strong objections to hiring the person. If so, I first make sure I understand the objections, then as long as I think they are supportable, I will reject that candidate.

Many organizations go further and eliminate anyone who does not get a strong endorsement from everyone. In other words, you only say yes to people that everyone on the team would want to work with every day. This strategy specifically bans lukewarm responses like: "This person does not fit here, but might fit somewhere else in the organization." While I understand that point of view, I will not automatically rule out someone who gets strong endorsements from most interviewers, but a lukewarm reception from one or two.

Testing Legalities

Before you give a candidate any test, make sure you know the laws that apply for your location and company.

Laws vary widely from country to country, and within the US, from state to state.

Usually I ask for an initial evaluation from each interviewer and based on that input decide whether the candidate is "hirable" or "not hirable." Then, when all interviews are complete, I ask the team to rank order the hirables. Depending on how many hirables you have and how many positions you are trying to fill, you can get more or less quantitative about the evaluation. However, in the end, I want to have consensus that we have chosen the best candidate or candidates.

Evaluating writing samples

Writing samples are critical to evaluating a writer.   Anyone who is serious about working for you as a technical writer will volunteer writing samples or have some ready to send you on request. If a candidate cannot come up with a writing sample, consider it a red flag. I have only hired one or two technical writers without a writing sample, and it was a mistake to do so. You might get lucky, but it is more likely you will find yourself saddled with a writer who cannot write.

Occasionally, you may run into someone who has worked for defense contractors or other companies that will not let him or her take home a sample. In that case you should get a writing sample as part of the interview (see above for some suggestions on how to do that).

I usually ask one person on the interview team to evaluate writing samples, though if you have got a lot of samples, you may need to spread the work around. Either way, I want the team to weigh the writing samples along with the interviews before we make a decision.

Here are some of the things I look for in a writing sample:

- ► **Writing:** Is the sample grammatical? Are there spelling and typographical errors? Is the writing clear and concise? Does the writer understand how to write different kinds of content, like procedural, conceptual, and reference?

- ► **Organization:** Is the sample organized logically? If the sample is in a traditional style, is the table of contents well organized? If the sample is written in a modular style, are the structure of individual modules and the connections between modules logical and consistent?

- ► **Understanding:** Can I understand what the writer is trying to tell me? And, does the writer show a good understanding of the subject matter? The latter may be more difficult to evaluate than the former, but you can often get a sense of how well a writer understands a topic by how clearly he or she describes it.

- ► **Style:** Is the style clear and consistent within any one sample? Writers often work to a style guide that constrains their writing. Therefore, I do not question particular stylistic choices, unless I have reason to believe they reflect the writer's style rather than a style guide. Rather, I look for a concise, clear, and consistent style.

- ► **Presentation:** How is the sample presented to you? The best candidates will gather a portfolio of samples with an explanation of each project, its objectives, and the results. A well presented portfolio tells me that the candidate cares about his or her work and understands the importance of context and preparation. A pile of random writings dumped on my desk tells another story.

Checking references

Always ask for references, and always check them. By some accounts, as many as 40% of résumés contain false information. Verify the claims of anyone you plan to hire. Managers often omit this step. I have been asked by many former employees to be a reference, and I usually ask them to tell me when they give my name to a prospective employer. I have not kept count, but I doubt I have received calls from even half of those prospective employers.

Given how easy it is to pick up the phone and call a reference, I am surprised at how infrequently it is done. When you interview anyone, ask for at least two professional references and one personal reference. Ideally, I like to have at least one reference who is one of the candidate's former managers, and if I am hiring a manager, I like to have a reference who has reported to the candidate.

Normally, you should check references after the interview and before you make a job offer. This means you only need to check references for candidates you consider hirable. Also, by waiting, you will go into the interview with fewer pre-conceptions.

As with the interview, you should only ask questions that are relevant and legal. Here are some typical questions. Some of these are only relevant to professional references and others are only relevant to personal references, but the distinction should be obvious.

- ► What dates did the candidate work at the company?
- ► Why did he or she leave?
- ► What was his or her salary?
- ► In what capacity do you know the candidate?
- ► What were his or her job title and responsibilities?
- ► What types of content did he or she work on?
- ► What were his or her most important accomplishments?
- ► What were the candidates strong and weak points?
- ► What can you say about the candidates work habits?
- ► Would you rehire him or her?

You may also want to ask questions related to the interview. For example, if you are not sure how big a role a candidate had in a specific project, ask the reference about that project. Since you almost always come out of an interview wishing you had been able to probe more deeply into one or more areas, you may want to explore them with a reference (of course, if you are really concerned, call the candidate and ask the question directly).

You may not get answers to all of your questions. Many companies instruct their managers to only give "name, rank, and serial number" answers that usually boil down to: dates of employment, and with the employee's permission, salary information. But, many references will give you more information, either through the way they answer the questions, or if they really like the candidate, more directly. That does not mean there is a problem with a candidate if you only get "name, rank, and serial number," but it is a nice plus if you get an endorsement from someone who has been told not to comment.

If you are working with a headhunter or contract house, they may have checked references ahead of time, but you should still do so yourself. Headhunters want you to hire their candidates, so they may not aggressively question a reference or probe beyond the basics. Make sure you get the information you need directly.

In addition to checking references, many companies do a background check to verify employment, education, and other factual statements on the résumé.   I think this is a very good idea; if your company does not do a background check, have a talk with your manager or Human Resources and see if you can get them to adopt this procedure.

Using Contractors and Contract Services

Most companies use a mix of three kinds of workers:

- ► **Regular Employees:** As the term implies, these are people who are employed by your company and, in the U.S., receive a W2 every year.

- ► **Project/Agency Contractors:** These are people who are brought in on contract from another company, typically to work on one particular project or group of projects.

- ► **Contract Services:** These are services, like translation, that usually take place off-site and often are provided by people who you never directly work with. Translation, editing, indexing, documentation conversion, and formatting, are all services that might be provided in this way.

There is a fourth category, consultants, who are usually brought in to provide advice, rather than to work on a project directly. For example, you might bring in an Information Architect to help you design your documentation structure. That person may create that structure, and deliver documentation about it, but he or she will not be involved in the content you create using that structure.

Each type of resource has its pluses and minuses. For teams larger than a few people, you will probably need all three. This section looks at the three types of workers and the advantages and disadvantages of each.

However, be forewarned that in many companies, you will have little control over the types of resources available. Some companies prefer regular employees, some prefer on-site contractors, and others prefer to outsource as much as possible. In those cases, you may have to make do with what you get, but if you understand the advantages and disadvantages of each, you may be able to build a business case for the type of resources you need.

Regular employees

For core tasks, regular employees are your best choice. A regular employee should have a stronger commitment to the company (including a possible a stake in the company through retirement and stock plans). In addition, you are more likely to get a long term commitment, which means you get greater leverage from training and greater productivity as the employee gains experience with the culture and practices of your company.

Ideally, if a job is critical to the quality of your product or service, and there is enough work to justify a full-time resource, you should assign that job to a regular employee. However, I find that many companies are moving to contractors, and contract services, including off-shore services, for technical writers. As a result, you may need to build a business case to justify hiring an employee over a contractor. the section titled "Managing the Hiring Process" (p. 53) discusses strategies for improving your hiring leverage, including building a business case.

Consider using regular employees in the following circumstances:

- ► **Long term projects:** A project that will last for three years would be a poor choice for a contractor you will need to replace in twelve months. Also, long term projects are more likely to be core projects critical to the company, and therefore, better served by regular employees

- ► **Critical content:** While a contractor can do work that is just as accurate and precise as an employee, I would want an employee working on critical content like the documentation for medical equipment or an aircraft. The learning curve is long enough and the need for accountability strong enough that you want committed, long term employees handling this kind of content.

- ► **Customer-facing activities:** If you host a Wiki, a forum, training, or other customer facing activity, you should consider using regular employees. I want people in those roles who are more likely to be around for a while and to be committed to the company. Customers get to know the people in these roles and build a rapport. It hurts your credibility if you change the people in these roles too frequently.

Contractors

At one time, at least in the U.S., it was not unusual to have contractors who spent many years with the same company, and who in many ways were indistinguishable from employees. It was not until I became a manager that I discovered that some of the people I had been working with were contractors rather than employees; to fellow workers, there was no difference.

This changed as a result of several lawsuits in the 1990s, most notably *Vizcaino versus Microsoft*. The plaintiffs in that lawsuit, who had been long term contractors with Microsoft, claimed they were "common law" employees of Microsoft and therefore entitled to benefits like discounted stock programs. After more than ten years of legal wrangling, Microsoft settled for over $90 million.

As a result, you might have expected companies to restrict their use of contractors in favor of regular employees. However, instead, many have increased their use of contract employees, and simply applied restrictions on contract employment to avoid the problems Microsoft had. Common restrictions include limiting the length of contracts, often to no more than one year, and restricting how long a contractor must wait before taking another contract.

The practical result is that you can only use contract employees for limited periods of time, and you have less freedom in how you can work with them. For example, you will most likely be unable to invite them to company functions, even ones held during working hours; to reward them, other than through their normal salary; or to punish them, except by terminating their contract.

On the other hand, you have greater freedom in hiring and firing, and your company does not need to provide benefits. While the hourly cost can be significantly higher for a contractor, the savings in benefits and greater flexibility make contractors attractive to companies.

Consider using contractors in the following circumstances:

- ► **Short term projects:** Contractors are a good choice for a short term project. Just be sure that the project will not require long-term maintenance. Otherwise, you may be better off with an employee.

- ► **One-off projects:** It is not unusual to be asked to handle one-off, projects, often in areas not related to your core work. Hiring a contractor can isolate the work from your team and minimize distractions.

- ► **Crunch time:** This one can be tricky; as Fred Brooks said in his classic book, *The Mythical Man-Month*,[7] "Adding manpower to a late software project makes it later." Even though Brooks was talking about software projects, his words hold true for any project.

  However, if you know you will need extra staff to help handle a short term increase in work, bringing in contractors makes sense. Just be sure you select tasks that can be handled with a minimum learning curve and include the learning curve in your schedule.

- ► **Non-core tasks:** Consider contractors for tasks like editing, formatting, and tools support. These tasks generally require less domain specific knowledge than content creation, and therefore, can tolerate more frequent turnover. Given the choice, I would prefer to staff these positions internally, but if you must use contractors for part of your team, this is a place to look.

Contract services

Also known as out-sourcing, contract services have always had a role in the teams I have managed. Companies regularly contract out janitorial, food, maintenance, and other non-core services. However, in the last ten years use of contract services and off-shoring has exploded. Many companies outsource anything that is not core to their business, including IT (Information Technology), HR, bookkeeping, marketing, and even documentation development.

Consider using contract services in the following circumstances:

- **Non-core tasks:** These are tasks that might distract your team from its main responsibilities. Graphic design and book design are examples. Every writer likes to think he or she is a good designer, and if left to their own devices, many will happily spend hours on graphics and book design.[1] While these tasks are important, you will be better off leaving them in the hands of professionals and keeping your writers focused on writing.

- **Specialized tasks:** These are tasks that require specific skills or knowledge. For example, most companies outsource translation. Translation requires specialized knowledge, which is best rendered by in-country translators. It also requires significant coordination, in large part because of geographic distribution. Because of this, even large companies outsource translation.

Other specialized tasks to consider outsourcing include: graphic design, tools development, web-site design and maintenance, and usability testing. As with non-core tasks, you need to evaluate your own situation, looking at both the talent on your team and how you can best direct that talent. For example, even if you have someone with usability testing experience on your team, it may be better to outsource usability test to keep that person focused on his or her core responsibilities.

---

[1]I know this from personal experience; had I not hired a graphic designer, I might still be redesigning this book's cover.

► **Process/technology shifts:** A major process or technology shift is daunting. If you are reorganizing your content, moving to new processes, or upgrading your technology, a good consultant can make a real difference. They bring two critical elements to the party: an outside perspective and a set of tools and techniques specific to their specialty. These two elements can help keep you on track and move through a transition more quickly and in the end less expensively.

► **One-time tasks:** These are tasks, like document conversion, that once completed, do not need to be repeated. Document conversion may end up being phased over several months, but once it is completed, there is no ongoing maintenance required. Because of this, and the specialized skills required, it is an excellent candidate for outsourcing.

► **Lower skill tasks:** These are tasks that members of your team could handle, but which are less expensive if outsourced. For example, you might find it more efficient to outsource copy-editing or document formatting. In addition, since these kinds of tasks are often cut to save expenses, you run a smaller risk of losing regular employees if they are handled outside.

In general, if a task would distract your team, require specialized expertise, or cost more if done in house, you should consider outsourcing.

Off-Shoring

Off-shoring is extreme outsourcing, that is, outsourcing to another country. Most of the circumstances that would lead you to consider outsourcing apply to off-shoring. The trade-off is between lower direct costs, because of the lower cost of labor, and higher indirect costs.

The direct cost of engineers in India, for example, can be half the cost of engineers in the U.S. However, the actual cost of off-shoring is considerably more than the direct cost. If you are considering off-shoring or are told you must off-shore some of your work, you need to factor in the indirect costs as well, including the following:

► **Coordination:** Unless the development team is co-located with the documentation team, there will be significant coordination overhead. The overhead will increase when the two locations are many time zones apart and staffed by people from different cultures, with different native languages.

Given the amount of contact writers need with the engineering team they support, I put co-location at the top of my list when considering off-shoring. I have seen organizations where the sole criteria for off-shoring the documentation was whether or not the product development was off-shored. They would not off-shore document-ation unless the product was off-shored.

- **Language differences:** Even if you off-shore to a country where your native language is widely spoken (for example, off-shoring English content to India), there will be differences in usage. While you may be able to communicate reasonably well with writers and managers, the odds are that the documentation will need to be heavily edited.

- **Turnover:** Because of the heavy competition for engineers, turnover can be very high. A recent report from the Hay Group says turnover in India is greater than 20% in high tech businesses.[21] I have seen anecdotal evidence that turnover as high as 50% is not unusual.

- **Rising costs:** The Hay Group report says that wages in India will rise by 14.4% in 2008, making it the fifth consecutive year of double-digit increases. The huge cost advantage that India had over the U.S. and other western countries has narrowed significantly.

Given rising direct costs and significant indirect costs, I would only consider off-shoring in two situations:

- Documentation for a project that is already off-shored, provided you can find people with the skills needed to write competent doc-umentation in your target language. The reduced coordination ex-pense from co-location, together with lower labor costs tip the scales here.

- Self-contained projects that can be cleanly separated from your core work. The less coordination required, the more likely that off-shoring will work.

In any other situation, I would avoid off-shoring for documentation work. Of course, you may not be given much choice. If that is the case, you may want to build a business case for keeping the project local, using the indirect costs as a justification. Beware that this can be difficult, since it necessarily requires some guessing. See Chapter 17, *Building a Business Case* (p. 169) for more information.

Hiring non-employee staff

Hiring contractors is much like hiring employees, you should still review résumés and hold interviews, but if you trust the contracting company, you may be able to delegate the background check and initial screening to them. Hiring contract services is also similar, except that you are hiring a company, rather than an individual, so your investigation will look at the company's track record.

Hiring a consultant is a different game. While a consultant will often have tangible deliverables, like a contractor or regular employee, those deliverables are more likely to be recommendations, rather than direct work products. If you think you need the services of a consultant, be absolutely clear about what you need, what the deliverables will be, and what the cost will be.

You also need to understand the connections between the consultant and the vendors of any technology he or she recommends. Some consultants make their living helping people choose technology, others make their living helping people implement a particular technology. If you are still selecting a technology and need help, make sure you get a consultant in the first category, one who is not tightly tied to a particular vendor.

For both consultants and contract services, speak with people who have worked with the company on projects similar to yours. As with hiring people, you can check on the web, both for what the company says and what others say about the company.

Managing the Hiring Process

One of the great frustrations of the corporate world is that hiring often works like a faucet that gets turned on and off randomly with no visible correlation to actual needs. In most of the larger corporations I have worked for, we would endure long dry spells where we could not hire, regardless of what work needed to get done, followed by a hiring binge that always seemed to occur just when the pressure eased. To further complicate matters, we were sometimes told to hire contractors and sometimes told to hire employees, but rarely given a choice.

This meant that in practice hiring was tied to a higher level budget, rather than any actual need. Even if you did not need to hire, when the opportunity arose, you did so anyway because you could not be sure when the next opportunity would arrive. The one good thing about this otherwise pathological behavior is that it can give you the luxury of training new hires, and it relieves you of the pressure to hire someone, anyone *right now!* The flip side, of course, is that you can find yourself with imbalances in staffing.

Gaining control of your hiring can be difficult. Even in the best companies, budget constraints can preempt the best laid plans. However, there are some strategies you can use to improve your flexibility:

- ► **Get the basics right:** Plan your team's work carefully, then execute to the plan. If you get the basics right, you will gain leverage for hiring and for anything else you need to do. If not, nothing else you do will help.

- ► **Be visible:** Technical writers tend to be introverts; we like to hide in our caves and write beautiful prose. That will not work for a manager; you need to advertise your accomplishments and the importance of your team at every opportunity.

- ► **Be part of a profit center:** Documentation groups often live in limbo between being a cost center (an organization that does not generate a direct profit) and a profit center (one that does). If your group is seen as a cost center, then the corporation will always be looking for ways to reduce your budget. If you are part of a profit center, they will be more open with their wallet.

- ► **Prepare a business case:** If you prepare a strong business case for hiring, you will be way ahead of the game. The people who control the purse strings understand business cases; you need to speak their language to communicate with them. Chapter 17, *Building a Business Case* (p. 169) discusses how to prepare a business case, and also talks about profit and cost centers.

In most companies, it will be easier to hire a contractor or use contract services than to hire a regular employee. If you are convinced you need to hire a regular employee, your success will depend on whether you convince your management that the task is a critical, long-term task or that the discipline is a *Core Competency* for your company.

A core competency is fundamental expertise, skill, or technology that differentiates a company in its market. Most companies do not consider technical communication to be a core competency. So, to justify hiring a regular employee, your best argument will usually be based on how the documentation contributes to the quality of a core product and how the long-term nature of the project justifies a regular employee.

The bias of most first-level managers is towards regular employees. The advantages of longevity and loyalty are strongly felt at this level, and in my experience it is more fun to manage regular employees (though to be fair, I have employed many outstanding contractors).

However, do not dismiss the usefulness of contractors, contract services, consultants, and even off-shoring. Each has its place.

6

Motivating

If you want to build a ship, don't drum up the men
to gather wood, divide the work, and give orders.
Instead, teach them to yearn for the vast and endless sea.
— Antoine de Saint-Exupéry

When I was a child I was in the Cub Scouts, and each year my Cub Scout den put on a show. I have no idea why we were doing shows when we should have been tying knots and pitching tents, but that was what we did. I had what I thought was a great idea for the show, and I spent days working out every detail, from the dialog to the casting.

On the day of our den meeting, I was so excited that I started my pitch to the den mother as I walked in the door. "Mrs. Smith, I've got a great idea for the show…" Before I was 10 words in, Mrs. Smith (as I will call her) stopped me cold, saying, "that's nice, but we've already got the show worked out." And, we went on to do the show she had planned, which incidentally I have no memory of, although I still remember my idea in excruciating detail.

I am willing to bet that everyone has had an experience like that, and if it did not make you feel small, unimportant, and demotivated then check your pulse. I knew, even as a small child, that this was not the way to motivate someone. I know now that Mrs. Smith was not trying to demotivate me; she just wanted to keep things moving.

What Does it Mean to Motivate?

Common wisdom is that managers are supposed to motivate. So, they try to motivate by dangling cheese, wielding whips, or my personal favorite, standing in the prow of the boat trying to look like George Washington. The problem is that while it is scarily easy to demotivate a person – even unwittingly, as the Cub Scout story shows – it is not possible to motivate someone else. In the real world people are moved by: *coercion,* which pushes from the outside, and *motivation,* which pushes from the inside. You can coerce, but you cannot motivate.

Coercion and motivation may lead you to the same results, but there are differences, both in what gets done and in how it gets done. As soon as any degree of coercion comes into play, most people turn off at least some critical faculties. After all, if I have to do it anyway, why should I spend a lot of effort figuring out whether it is the right thing to do or whether there are better ways to do it?

When someone says "Yes Sir!" to me, I get a chill. I know the job will get done, but if he or she knows a better way of doing the work, or sees something I have not noticed, I will not hear about it. I would rather hear, "have you considered this alternative?" or "are you sure we need to do this?" because that person is engaged and therefore motivated.

Coercion sets up a dependency between you as the "order giver" and the other person as the "order taker." Someone who is "just following orders" has relinquished his or her independence to the order giver. When a person relinquishes independence, he or she also relinquishes responsibility for the outcome.

This is so common that it has become a cliché. Mr. Dithers tells Dagwood to make a reservation at the fancy steak house for a big customer; Dagwood starts to tell him that the customer is a vegetarian, but Mr. Dithers doesn't listen and tells Dagwood to shut up and make the reservation, so he does and hilarity ensues. The sad thing is that this happens in the real world every day. I have seen otherwise intelligent people do very dumb things and accept no responsibility for the outcome, because "the boss told me to, and he knows what he wants."

Most people are able to manage their home lives just fine. They plan and execute their activities independently. They manage their resources intelligently, maintain their home and car, and invest for the future. They take responsibility for their lives and the lives of their families without anyone telling them what to do.

Then they go to work and turn into robots for 8 hours a day. At work, they move into a more or less coercive environment, which encourages them to turn off their brain and hand responsibility to the boss. Some of this is inevitable. Unless we are independently wealthy, all of us hand over a big chunk of our lives to earn the money we need to live. Even when it is a fair exchange, and as a manager you need to do your best to ensure that it is, there is an element of coercion that is unavoidable. Our job as managers is to acknowledge the necessary coercion and do all we can to drive unnecessary coercion out of the system.

Common Demotivators

Since managers do not directly motivate employees, we must instead give employees an environment that to the greatest extent possible gives them control and removes coercion and demotivators. Here are a few:

- **Not listening:** Pay attention to what people say. Then, if you are not going to take a suggestion, explain why. Not only does this show respect, if you are paying attention you might just learn something.

- **Being inflexible:** If you are paying attention, you might hear something that will change your mind. A natural inclination is to resist that change, especially since the idea came from someone else. Resist that inclination.

- **Imposing arbitrary schedules:** Writers should plan and schedule their own work, and managers should work with writers to accommodate the inevitable externally imposed schedule constraints and avoid adding more.

- **Taking away power:** Do you insist that writers be in the office at specific hours even when there is no reason to do so? Do you write your team's plans and schedules? Do you micro-manage? Do you insist that your team communicate with your manager only when you are in the loop? All of these things, and lots more like them, take away power and demotivate.

- ▶ **Withholding information:** Information is power, and withholding information is another way to take away power. Managers withhold information all the time. Often this is for good reasons, for example, to protect confidentiality. But, most non-confidential information, especially information about strategy and tactics – all the "why" information – should be shared. Besides confidential information, the only other kinds of information I do not normally share are rumors, unfounded opinions, and administrivia; information that just wastes time and does not add anything useful.

- ▶ **Disrespecting people:** Just because you are the boss does not mean that the people who work for you do not deserve respect. Keeping people waiting, belittling them, or making decisions for them are all disrespectful.

- ▶ **Not leading:** All this talk about coercion might make you think that managers should just sit back and let things happen. Not true. Having an empty suit as a manager is nearly as bad as having Simon Legree. You need to know where you want your team to go, and you need to communicate that with your team.

- ▶ **Personnel craziness:** Every corporation has its pathologies, especially with regard to personnel issues. There will be more on this in other sections, but for now, it is enough to recognize that corporations tend to think that personnel systems like employee ranking, performance evaluations, management by objectives, and this week's management fad are motivators, even though they rarely are.

- ▶ **Market realities:** If things are going badly for your company, or the particular product you support, that is a powerful demotivator. There is not a lot an individual manager can do in this situation, but that does not mean it will not have an impact on you and your team.

Removing Demotivators

The key to removing demotivators can be summed up in the words of W. Edwards Deming, "Drive out fear."[1] Necessary coercion – that which is built into the system or is part of leading – should not cause fear unless you are trying to correct a disciplinary problem. As soon as fear enters

---

[1]*Out of the Crisis* [12], p. 59. A must-read for managers.

the picture, it drives out rational thought and even smart people will do dumb things.

If the people who work for you are afraid to speak up, if they are afraid to point out when you are wrong, if they are afraid to express an opinion that is contrary to yours, then you have stepped over the line from necessary coercion to unnecessary coercion, in other words fear. While you may have a group that does your bidding, you will not have a group that is motivated to do its best work.

Building a Motivated Team

While removing demotivators is essential, that is not the full story. You need to make sure your team understands your objectives and priorities, then get out of the way and let them fill in the rest. Stay engaged, but let your team be the problem solvers.

Their role as problem solvers comprises more than just figuring out how to get from place A to place B. They also need the latitude to challenge your objectives and priorities and suggest new ones as appropriate. Maybe place B is the wrong destination and place C would be a better choice. By getting the team involved, they take ownership of both the destination and the tactics that will get you there, and they will be more motivated.

This is not easy to do. Your natural tendency will be to over-specify the objectives. In Chapter 3, *Power and Influence* (p. 18), I describe a project that included a web interface for engineers to enter information. The essential objective was to make it easy for engineers to enter information. The objective was *not* to build a web interface, and it certainly did not require a particular look and feel.

Despite that, I had strong opinions about what the web interface should look like and what features it should include, and I spent a lot of time talking about those opinions with the engineer who developed the page. In fact, I spent way too much time, and the web page took longer to develop than it should have.

The engineer understood he could meet the essential requirements quickly, and then refine the design as he got feedback from users. Once that fact penetrated my thick head, I got out of the way, and the site was set up quickly. It turned out to be a big win, and the things that I was so worried about got cleaned up quickly with little muss or fuss.

While my objective was sound, my mistake was defining the solution. Even if that solution had been optimal, which it was not, forcing a particular solution was counter-productive and demotivating.

Fortunately, I was saved because I had created an environment that minimized demotivators and encouraged the engineer to assert himself. In a more coercive environment, he might have said nothing and the project could have failed or at best taken much longer than it should have. A good environment is a safety cushion that can save you when you mess up.

If you can do these things, you have got a good shot at having a motivated team. However, no matter how good the environment, there will be elements out of your control. For example, you may be saddled with dysfunctional management or have an employee who has a talent for demotivating others.

In the first case, do your best to keep your team as sane as possible and buffer them from the craziness. Be an umbrella that deflects garbage from above. In the second case, unless you can convince the employee to change, you need to get him or her out of your environment. Group dynamics are delicate, and a bad apple really can do serious damage.

But, if you keep the environment un-coercive and supportive, if you define and communicate clear objectives that have been distilled to their essence, if you let the team plan how to reach those objectives, if you are a true advocate for your team with the rest of the company, and finally if you get out of their way and let them take responsibility for their work, you will be rewarded with a motivated team.

7

Managing Change

Change should be a friend.
It should happen by plan, not by accident.
— Philip Crosby

Users will change their habits when the pain of their current situation
is greater than their perceived pain of adopting a possible solution.
— Pip Coburn

When our daughter was a toddler, she suffered from repeated ear infections. The standard treatment at that time was Amoxicillin, an antibiotic dispensed as a thick, pink liquid that tasted awful. Our daughter hated the stuff and resisted taking the medicine, even though we told her it would make her feel better, which it did.

When her ear infections continued to recur, her doctor suggested putting tubes in her ear drums, a procedure that allows the ears to drain, making them less susceptible to infection. We chose to do this, and it solved the problem. Because our daughter was too young to understand what was happening, and because the doctor used an anesthetic, the procedure was at worst a bit uncomfortable for her, and unlike her parents, she suffered no anxiety about the procedure in advance.

However, consider if it was you, rather than a toddler. You probably would not think twice about taking a nasty tasting medicine if it would help. And, you would probably accept an injection in your arm, though maybe with a little anxiety. But, I suspect that you would react much more strongly to the prospect of putting a tube through your ear drum. You would question the doctor about the efficacy of the procedure and look for a way to get relief without enduring the perceived pain of putting a tube through your ear drum. Your perception of the procedure and that of the toddler are almost completely reversed, even though we are talking about exactly the same condition and treatment.

Technology changes in the work environment are a lot like ear infections. You complain about the pain, maybe impaired productivity, and when someone suggests a way to fix the pain, maybe a new technology, you weigh the potential benefits versus the perceived pain of adopting the fix. If it is the equivalent of taking medicine or getting an injection, you go forward; if it is the equivalent of punching a hole in your ear drum, you probably resist. The problem, of course, is that some of the people affected by the change will see it as taking medicine and others will see it as punching a hole in their ear drum.

In this chapter, I look at how you as a manager can facilitate change and make productive change a part of your environment. This is a tough one; the most critical mistakes I have made as a manager occurred when I was unable to bring about a needed change or sustain that change until it became ingrained in the organization. Regardless of how brilliant your proposed change is or how important a particular change may be to the success of your organization, if you cannot gain acceptance of the change or cannot sustain it, you will fail.

The Burning Platform

Common wisdom is that you need a "Burning Platform" to cause change. A burning platform describes a crisis so painful that you are forced to act. The classic image is that of an offshore oil rig on fire. The prospect of jumping into the ocean is so perilous that the workers will not jump until it is clear that remaining on the rig means certain death. Jack Welch, former CEO of General Electric, is famous for using the burning platform as part of his transformation of GE. While he saw clearly the problems he needed to address, he needed the rest of GE to see the same things he saw. His ability to communicate that perception was part of what made him successful in transforming GE.

While the burning platform is effective, using it is an admission of failure. There are situations, and the GE that existed when Jack Welch became CEO is arguably one of them, where you have no choice; you need to point out the flames and use them as an incentive to take strong action. But, if your team will not make a change unless the flames are licking their feet, then you have two crises on your hands; the flames and a team that has not embraced change as part of its normal day to day work.

I am suspicious of the burning platform. I have seen managers try to generate a burning platform out of whole cloth. As you might guess, this is pretty much guaranteed to backfire. People may disagree about the severity of a problem, but they will usually agree on whether the problem exists. If you try to manufacture a problem, your team will know, and in addition to deciding you are a moron, they will resist change.

The other place where a burning platform will backfire is when you are standing on the burning platform, but your team is happily treading water near shore. Unless the people who need to change feel the flames licking at *their* feet, they will not be inclined to move.

If you do have a true burning platform, it can be a catalyst for change, but only if it really is a legitimate burning platform, and only if the crisis affects the people who need to change.

The Change Function

A more nuanced, and to me more compelling, view of change is described in Pip Coburn's book, *The Change Function*[9]. Coburn argues that change is a function of "perceived crisis," how bad you think things are, and "total perceived pain of adoption (TPPA)," how hard you think it will be to change. Regardless of how compelling a technology may be on the surface, it will not be adopted unless people believe that the "pain" of changing is less than the "pain" of remaining in their current situation.

This explains why some technologies, as cool as they may be on the surface, are never widely adopted. The classic example is Picturephone, which does not address a compelling "crisis" for most people, and which has a high perceived pain of adoption. Another example is the Segway, a very cool technology that does not address a strong enough crisis to overcome its TPPA.

An important aspect of the "change function" is that it is a function of the potential *user's* perception. For example, consider RFID (Radio Frequency Identification). RFID is a technology that performs the functions of bar codes without the need to directly scan information. Instead of a bar code reader, which needs to be positioned where it can visually read the tag, you can read an RFID tag at a distance using a specialized radio receiver.

If you live in or near a big city, you have probably seen toll roads or bridges that use RFID technology with names like EZPass, Fast Lane, or Smart Tag. If you have the right device, you can drive through an RFID equipped toll booth without stopping; the system registers your passing by reading the RFID device and debits your account for the toll. RFID has been a big success in this application. It provides a clear benefit to the user (speed and in some places a discounted toll), and a clear benefit to the vendor (fewer delays, fewer toll booth attendants, and often a chunk of pre-paid cash).

RFID is also used to tag merchandise. Big retailers like Walmart like RFID because it speeds up the handling of merchandise and gives them more control over inventory. However, to do this, their suppliers need to also adopt RFID, at considerable cost. Since few of the benefits accrue to these suppliers, there has been at best lukewarm adoption of RFID, much of which can be attributed to arm twisting by retailers who have the clout to force their suppliers to use it. When suppliers do not perceive a benefit that is worth the pain of adoption, they resist the change.

Many of the highly touted documentation technologies, such as content reuse, XML, and Content Management Systems, promise to relieve one or more aspects of a *corporation's* perceived crisis. Given that these documentation technologies provide significant benefit to the corporation – an assumption that can and should be challenged *before* adoption – they also need to provide significant benefits to the actual users or their adoption will be troubled.

The perceived crisis and perceived pain of adoption may change over time. For example, if the cost for RFID gets cheap enough for suppliers, the TPPA for RFID will go down. Or, the cost of Picturephone service might go down significantly as more and more people buy cell phones with cameras in them. In either case, the rate of adoption for these technologies could go up.

Leading Change

I have spent years trying to move organizations towards better structured and more productive environments. Some of those efforts have been successful, others have failed. Along the way I have picked up some hard lessons that may be useful.

Probably most important is that people respond best to concrete, personal and urgent challenges. If your house is burning, you will act and act quickly. If it is not burning right now, you are less likely to promptly fix wiring problems, replace batteries in smoke detectors, buy a fire extinguisher, or take other simple preventative actions. This is one reason why the burning platform is effective; it is concrete, personal, and urgent. As you look at the suggestions below, consider that any of them will only be effective to the extent that they address concrete, personal, and urgent concerns.

Let's look at four critical things you need to consider as you face change: the perceived crisis, the desired future state, the perceived pain of adoption, and your team's attitude towards change. Most of what follows is written in the context of technological change, especially the adoption of new technologies. However, the concepts apply to other kinds of change, too.

Perceived crisis

As manager, you must have a clear understanding of the reasons for change. If you do not, stop right here and make sure you understand and can articulate why it is necessary to change.

Your job in communicating the perceived crisis is to make the current situation and the reasons for change concrete, personal, and urgent to everyone who is affected. In addition to your team, this includes your management team and any other team that is affected.

Here is where the notion of a Burning Platform may be useful, especially if you have a true crisis on your hands. Do not be shy about outlining the dangers in your current situation and the importance of changing, but at the same time stay real; your motivation for change needs to withstand scrutiny and make sense to those you are trying to convince. If you misrepresent the situation or artificially create a crisis, they will figure it out sooner or later.

Make sure as well that you portray the current situation in ways that apply directly to *everyone* you need to persuade. I have been in situations where I made a compelling argument to writers for a change, but neglected to make an equally compelling argument to management, with the result that management saw no reason to support the change. Everyone with a stake in a change needs to have a reason for supporting it.

The perceived crisis is half of Coburn's "Change Function," and therefore is critical to leading change. But, it is very easy to forget about the perceived crisis and skip to the perceived benefit, especially with new technology. You need to separate real need for change from the "Gee whiz, wouldn't it be nice" glow you get from cool technology and the salespeople who push it. If you cannot identify a real need for change, beware.

Desired future state

The desired future state defines the point you are aiming for. As with perceived crisis, you need to articulate the desired future state clearly and specifically for each constituency. If you can agree on the perceived crisis and the desired future state, you are halfway to making change happen.

However, it is easy to get off track. The most common, and very dangerous, mistake is to confuse the desired future state with a specific means for reaching that state. If you start discussing the future state in terms of specific technologies or methodologies, the chances are that you are making this mistake. The danger here is that humans love to solve problems and will happily start proposing solutions before they are sure what the problem is. Therefore, make sure you define the desired future state, in terms that pose as few constraints as possible on the solution, before you dive into problem solving.

Once you have a common understanding of where you are and where you need to go, leading change will become easier because you will have a framework for engaging your team and other affected parties in defining the details of a solution.

Pain of adoption

Managing technical change includes picking the right technical solution, which is discussed in Chapter 16, *Acquiring Technology* (p. 157), and deploying that solution, which is discussed here. In many ways, the most important detail in planning a solution is deployment. Paradoxically, the deployment of a solution can be more important to adoption than the solution itself and is often more difficult to pull off. A good deployment is one that minimizes the Total Perceived Pain of Adoption for your solution. No matter how good your solution is, if the TPPA is too high, it will not be adopted except by sheer force.

Beyond the technical details of deployment, there are three factors that you need to consider as you roll out technology: phasing, participation, and training. Let's look at these in turn:

- ► **Phasing:** A solution that requires a major one-time shift in processes or tools will usually have a higher perceived pain of adoption than a solution that can be phased in over a period of time in smaller steps.

  It will almost always be better to deploy a solution in phases, but there are some potential dangers. It is easy to lose momentum, either with your team or with management. The latter is especially dangerous because corporations live in the short term, and it is distressingly common for new management to torpedo the projects of their predecessors. At the same time, trying to make too big a change at one time can stop you dead in your tracks before the project has gone anywhere.

  Here are some of the keys to keeping momentum going:

  - Make sure each phase delivers benefits to your team and management.

- Front load expenses as much as possible so you do not need to ask for money separately for each phase. If you cannot do that, try to make each phase cheaper than the previous one.

- Front load the user benefits. Once users see the benefits to them, they will become advocates for subsequent changes. This also gives you good news to communicate to management.

- Make sure each phase stands on its own. That way, if you are cut off in mid-stream, you will have a workable environment.

- Communicate progress widely and frequently. Make sure users and management are frequently reminded about the great things you have already accomplished, the greater things about to happen, and the amazingly wonderful future that is just around the corner.

▶ **Participation:** Part of keeping momentum going is keeping participation active and positive. On one project, where I managed both the engineers developing the tools and the people using the tools, I let users attend the developer's meetings. While there were some topics they did not care about, having them there kept their issues in front of the developers. As a result, the development was in many ways led by the users, who became committed to the project and then helped recruit and train other users.

The other critical part of participation is keeping management informed with frequent updates. This is a case where "out of sight" really means "out of mind." If you do not keep your managers informed, they will get their information from other sources, most of whom do not have the same investment in success that you do. If you are unlucky, they may get information from people who would be happy to see you fail. While it may be hard to keep naysayers away from your management, you can make sure you are first with both good and bad news. Also, make sure you give demos frequently, so everyone can get a tangible sense of what you are doing.

▶ **Training:** Every time I have proposed a technology change, the first question from my team is almost always, "how are you going to train us?" Resistance to change is inversely related to the amount and quality of training available to users. Do not ignore training or offer it half-heartedly. Instead, plan for it from the beginning and factor in the cost, both the cost in dollars and the cost in the time your users will need to learn the system.

On the project mentioned in the previous section, the developers trained the first users, who then trained later users. In the right situation this will build a strong, committed set of users. However, most of the time you need to get people working quickly and cannot afford to have them spending time training other users. Therefore it is probably best to plan for formal training from the vendor or a skilled third party.

Attitudes towards change

Most of what I have discussed in the previous sections has to do with managing the introduction of new technology. Of course, as a manager, you need to deal with many other kinds of change as well, some of it planned and some of it unplanned. I have found that the best strategy for dealing with change in this general sense is to follow the advice of Philip Crosby in the epigraph to this chapter, "Change should be a friend. It should happen by plan, not by accident."

The best way to do this is to make change part of your environment. You need to constantly look for ways to improve your product and your productivity, and you need to constant implement improvements as you see them. A team that is continuously looking for and implementing improvements will see potential problems and avert them long before they become burning platforms. They will also learn to embrace change as normal and not an aberration.

I use several techniques to keep change "in the air." One of my favorite techniques is to regularly float crazy, and sometimes not so crazy, visions of the future. The Release Note project discussed in Chapter 3, *Power and Influence* (p. 18) is an example of this. I probably drove the Release Note author crazy dropping by and brainstorming about a future where the release notes were kept in a database as modules and assembled automatically for publication. At the time, we did not have any particular problem, but we did have room, as you always do, for improvement. I did not force a change, I just painted a picture of the future. The writer grabbed this vague idea and ran with it, coming up with a structure that took the first step towards the change discussed in that chapter.

Another way to keep change in the air is to constantly question what you are doing. Is your current structure the best it can be? Are your interactions with engineering as smooth as they should be? Is there a more effective or efficient way to create deliverables? What would your documentation look like if you were not limited by tools, techniques, or schedules? If you find limitations, how can you remove them? By regularly asking these kinds of questions, you engage your team in a constant dialogue focused on continuous improvement and change.

Your own attitude towards change is critical. Keep your mind open when members of your team or management propose change. If you do not take them seriously, they will not take you seriously. If you have concerns about a proposed change, use the suggestions in this section to help clarify everyone's understanding of the current situation, proposed future state, and the means for getting from the former to the latter. Even when the proposed change has been mandated by management, and you have no choice but to implement it, you can still take the time to understand the motivation and communicate it in the best possible light to your team.

In the end, no matter what you do, you will constantly face change, some welcome, some not. Having both a good attitude and a strategy for dealing with change will put you ahead of the curve. You will see the need for change and take action before change gets forced on you, you will have the tools needed to make the best of unwelcome change, and you will keep your organization happier and more productive.

8

Employee Performance Evaluation

We judge ourselves by what we feel capable of doing,
while others judge us by what we have already done.
— Henry Wadsworth Longfellow

Unless you work at a very small company, the odds are you will give and receive yearly Performance Evaluations (PE). I have mixed feelings about this ritual. At its best, the annual PE gives you the opportunity to gather input on each employee's job performance, assess the previous year's accomplishments, and make recommendations for development. It focuses your attention, and the employee's, on things you may not pay attention to the rest of the year. At its worst, it gives you the opportunity to show your ignorance of your employees and gives your company the opportunity to force fit them into arbitrary ratings.

The best performance evaluations I have received were those that provided insight into my strengths and weaknesses, and gave me one or two specific ideas I could use. For example, for years I was rightly criticized for cutting deadlines too close. By nature I work best under pressure, and I tend to be most productive when I have a deadline looming. Over the years, my "just in time," approach to deadlines added unnecessary tension and risk to the projects I worked on.

Pointing that out to me year after year, which is what a whole string of managers did, was not useful. The manager who helped me finally make progress went further. He helped me understand the impact I was having on other members of the team, and he helped me come up with some useful strategies for improvement. One of those strategies is embodied in the way this book was written. I placed early versions of several chapters on line, with a commitment to provide a regular stream of new content. In effect, I created a series of short term goals, each of which could be reached "just in time" without having an adverse impact on the ultimate result.

That is the PE at its best, but there is another, darker side. While most companies tout the value of the PE for professional development, they really use the PE to sort employees into winners and losers; those who will get raises and promotions, those who will remain in the same place, and those who will be shown the door. Professional development usually comes in a distant second.

This reality changes the character of the PE for both manager and employee, robbing it of much of its potential value. For many employees, the PE ritual is simply a game where you puff up your accomplishments to maximize your chance at the goodies. And, for many managers, it is just another empty exercise in pushing paper that you try to get through with the least possible effort.

Given this reality, and the time pressure all managers live with, you can bet that most will cut corners; I have done it and I do not know anyone who has not. But, after cutting the corners for way too long, I realized that this was one area that actually deserved my attention, and I changed my attitude. What I realized was that despite the distortions that companies place on the process, the PE ritual still provides unique value for both the manager and the employee.

I also realized that, like planning, the PE process is more important than the PE document. The process of getting feedback, analyzing it, formulating recommendations, and talking one-on-one with each member of your team provides benefits you cannot get any other way. To make the process work, you need to approach the effort seriously, invest some time, and combat the demotivators.

The Ritual

According to Merriam-Webster, a ritual is "an act or series of acts regularly repeated in a set precise manner." While the details vary, the core of the PE process fits that definition. In fact, the ritual is so consistent that you can buy software that will lead you through the entire process. I have seen packages that will suggest wording for common situations, give you categories and canned responses, and warn you when you use a word that might offend someone. As you might guess, I am not a fan of these programs; if you manage writers, you should be able to write a PE without help from a program.

Here are the main elements of the PE ritual:

- The manager gathers input from the employee's peers, other managers, and if applicable, anyone who reports to the employee. Sometimes the manager will also ask the employee for a self-assessment or a summary of accomplishments.

- The manager writes the PE, almost always using a standard form. Sometimes the forms are loose, other times they are very tight (for example, Employee has: 1) Excellent interpersonal skills, 2) Very good interpersonal skills, 3) Good interpersonal skills, …).

- The manager, his or her peer managers, and their manager meet to sort the winners and losers and assign a rating or ranking to each employee. This happens at several levels in the company, merging at each point up the ladder.

- The manager and employee meet to discuss the PE and the rating or ranking.

- The employee adds comments to the written PE, negotiates changes with the manager, and signs it.

- The manager files the PE with Human Resources, never to be seen again.

Let's take a look at the process in detail.

Gathering Input

Gathering input is relatively straightforward, but critically important. Here are a few guidelines for getting the most out of this part of the process:

- ► **Spread a wide net:** Get input from everyone who has worked with the employee. I also ask for input from everyone else on the same writing team, even those who are not working directly with the employee.

- ► **Include the employee:** One of my managers asked me to write my entire PE; he glanced at it, said it looked fine, and signed it. I would not recommend going quite this far, but you should ask employees for their input up front. Usually I have them give me an assessment of their accomplishments over the last year. This gives them a chance to shine, and it makes sure you do not miss anything important. Nothing makes you look more like the pointy headed boss than forgetting an important accomplishment.

- ► **Be flexible:** Even in companies that live and die on email, each person has a preference for communicating sensitive information. I generally send an email request, but state that I will take input by email, phone, or in person.

- ► **Keep feedback confidential:** Be clear about confidentiality. I always state that comments will be confidential unless the person giving comments explicitly says it is okay to reveal them. And I keep them confidential. If I use a particular comment in a written PE, and I often do, I make sure that the comment is unattributed and does not reveal itself in the wording.

- ► **Ask open-ended questions:** Use open-ended questions like, "Tell me about Martha's interactions with your team," rather than close-ended questions like, "Did Martha work well with your team?" A close-ended question pushes the response towards a yes or no answer while an open-ended question invites a deeper discussion. While it is fine to ask a few close-ended questions where you really want yes or no answers, make sure you include some open-ended questions, including a catch-all question like "What other feedback do you have for Felix?" to make sure you do not close off the discussion prematurely.

► **Ask positive questions:** I usually probe for "weaknesses" using questions that ask for positive suggestions. For example, a question like, "What could Rachel do to improve her interactions with your group?" will get you a more useful answer than, "What are Rachel's weaknesses?" Most people will not directly answer the second question unless there is a big problem, but most will be happy to answer the first with some useful ideas.

► **Follow up:** If you do not get a response or you have a question, call and ask. While I will not hound someone, I do like to get a response from everyone I ask; often it is the reluctant ones who give you the most revealing input.

► **Thank everyone:** I always send a thank you email to everyone who replies, even if the reply was, "I did not work with Mary this year and have no input." This is a yearly ritual, and you will be back asking for input next year. In the same vein, I always respond when asked for input by other managers.

Once I have collected input, I open up a file in an editor and gather everything together in one place. I then read it over looking for common themes and surprises. I also look for a few choice quotes to use in the written evaluation. There is nothing quite like a glowing quote from a peer to brighten up a PE. Just make sure that the quote does not reveal the author unless you have permission.

Writing the Evaluation

Once you have the input and have gathered your thoughts, identify one or two themes. Like any other presentation, you will at best be able to make just a few points stick. If there is some obvious area that needs to be worked on, that is a theme. If not, look for areas of strength that could be developed.

While most evaluations focus on "Areas for Improvement," I think you should look at strengths first. According to Marcus Buckingham and Donald Clifton's *Now, Discover Your Strengths* [8], "Each person's greatest room for growth is in the areas of his or her greatest strength." In their view, companies spend too much time trying to "fix" weaknesses and not enough time trying to capitalize on strengths.

While some weaknesses must be addressed – chronic bad work habits are a good example – most people will get more benefit if they improve in the area of their strengths and simply find ways to work around their weaknesses.

Once you have found your themes, you need to find a way to work them into whatever format your company has foisted on you. If you are un-lucky, you will need to fill in a bunch of tightly constructed forms with radio button ratings: "Employee's interactions with co-workers are: Always excellent, Always good, Usually good, Sometimes good, In need of improvement." This type of format makes it difficult to develop useful themes, but even the worst of them should give you space for free-form comments. Focus your effort there.

Here is how I write the evaluation:

- ► **Review the form:** Most forms use reasonably standard categories with titles like Interpersonal Interactions, Job Skills, Leadership Skills, and so forth. I look over the categories and consider how each one aligns with the themes. For now, I ignore ratings, radio buttons, or any other quantitative measure.

- ► **Identify actions:** For each category, I identify some suggested action that supports the theme. Examples include: taking courses, taking on new assignments, changing responsibilities, or working around weaknesses. Sometimes they are directives rather than suggestions, but I prefer suggestions wherever possible.

- ► **Select quotes:** For each category, I select at least one quote from the feedback that reinforces the theme.

- ► **Write individual sections:** Using the themes, quotes, and sugges-tions, I write a short section for each category. A typical paragraph for a writer whose strength is consensus building might look like the following:

> **Leadership Skills:** Imogene is an effective leader who knows how to build a strong consensus. One colleague said, "I thought we would never come to agreement on the structure of the new User's Guide, but Imogene was able to get a group of eight engineers from three different groups to

sign off on a plan in just two short meetings." I suggest that Imogene further develop this strength by taking the course, "Advanced Cat Herding: Bringing Contentious Teams to Consensus."

► **Write a summary:** Summarize and reiterate your themes. There is almost always a place for a summation at the end of the PE. I use this space to re-iterate the theme and the most important suggestions. I often drop in another quote or two.

► **Assign ratings:** If I need to assign ratings, I will do that at this point. However, if I am doing evaluations for a group of people at the same time, I will come back after I have finished all the write-ups and review the ratings for consistency.

► **Review with employee:** Give a copy to the employee to review. You may need to withhold some information, like ratings, since ratings are usually reviewed by management before they are revealed to the employee. However, with that one exception, I share the share the complete PE with the employee before the formal discussion. Since you are dealing with professional writers, they are sure to have some improvements, which I always take unless they change the meaning in a way I cannot support.

If you have an employee who has some serious issues that need to be addressed, use the same approach. Use quotes, your theme, and strong suggestions or directives. You owe it to the employee to be honest, and you need to document your actions clearly in case you need to take corrective action. About the only time anyone other than you and the employee will read a PE is when they are trying to determine if you followed the correct procedure for terminating employment, so write with that possibility in mind.

You will also probably need to deal with objectives. I think it is important for everyone to document a set of professional objectives. Professional objectives are things like: "Become an expert in using Photoshop" or "Publish an article about subject X in a professional journal." I like to have these kinds of objectives in a PE because they formalize what might otherwise be vague desires.

Some companies take this further and incorporate job objectives like: "Complete the User's Manual by October 17." While this is clearly not a substitute for your project planning documents, stating objectives in the PE clarifies responsibility and gives you a starting place for next year's discussion.

The main drawback to including job objectives is that very few job objectives, and even fewer schedules, remain constant over the course of a year, so when you review this year's objectives next year, they may bear no relationship to what actually happened. As long as everyone understands that objectives change, that is no big deal. Therefore, I think the advantages of documenting job objectives in the PE outweigh this drawback. It does not hurt, however, to use relative wording like: "Complete the User's Manual on schedule to ship with the product." That way, if the product slips, the objective is still valid.

Once you have completed the written PE, let it "age" for a day or two, then re-read it. This is especially important if you are sending a strong negative message, but I try to do this for every PE. I guarantee you will improve at least some wording and every once in a while you will stop yourself from something very embarrassing. I once put an entire section from one person's PE into another's by mistake and only caught the error in that last re-read.

The Employee Discussion

The shortest PE discussion I personally know about was given to a colleague while he was standing at a urinal in the men's room. His boss came in, unzipped at the next urinal over, and said, "You haven't had your PE yet, have you? You are doing a fine job." With that, he zipped up and left. At least it was a positive appraisal, but I would not recommend this method.

The PE discussion is the heart of the process. Of course it is important that you gather input and write a PE, and as a manager you will learn a lot just from doing those two things, but until you have communicated with your employees and gotten their responses, you are not done.

The discussion is also the most perilous part of the process. Even though I have been doing PEs for years, I still cannot accurately predict how an individual will react to his or her review. For example, I have broken the "never surprise" rule a few times, and that usually leads to an angry response. But, sometimes the employee takes the surprise calmly and positively.

On the other hand, I remember one employee who became unhinged by what I thought was an innocuous suggestion to re-read Strunk and White. Everyone should read William Strunk and E.B. White's *The Elements of Style*[36] at least once a year, but that did not stop this particular employee from taking umbrage.

The keys to a successful PE discussion are preparation and attitude. Being prepared means that you have gathered useful input, distilled that input into one or two clear themes, thought about – and if necessary practiced – how you will deliver your message, and considered the likely impact of your feedback.

Attitude means that after all that preparation, you are still open to input from the employee and you are ready to listen to what he or she has to say. In combination, these two factors make it possible for you to say what you need to say precisely and concisely, then devote the rest of the meeting to the discussion.

Preparing for the discussion

Here are some guidelines for your preparation:

► Give the employee a copy of the written PE a few days ahead of time so he or she has plenty of time to read it. If I am going to discuss a rating or ranking, I will leave that off the advance copy, but otherwise I provide a full copy.

► Pick out a private place for the discussion where you will not be interrupted. If you have a private office, that is a good place, but make sure you will not be interrupted.

► Schedule more time than you think you will need. I never schedule less than an hour, and I prefer not to schedule in front of a hard commitment.

► Re-read the PE to refresh your memory about your main themes.

► If you are delivering a strong negative message, especially if you have any reason to believe you may have to terminate someone's employment unless he or she improves, get some suggestions from your Human Resources representative on how to handle the situation. There are also some ideas in the section titled "Handling difficult situations" (p. 83).

Unless I have a specific negative message that must be delivered, I try to allow the employee to set the direction of the discussion. After all, my main points are in the written PE. While I will re-iterate the most important points to make sure they are understood, I invariably find that the greatest value in the discussion is from what the employee has to say, not what I have to say.

Discussing objectives

A good discussion will begin with a look at results from the last year, focusing on what can be learned from them. It will then move to a discussion of objectives for the next year. As mentioned earlier, you need to consider two kinds of objectives, work and professional. Work objectives are job assignments; professional objectives are personal objectives aimed at professional development.

I generally try not to dwell on job assignments in the PE discussion. It is very easy to lapse into a discussion about project X or project Y when you should be talking about job performance. While job objectives are important and need to be discussed, there are opportunities to discuss them outside the PE process. Therefore, in the PE discussion, I use job objectives primarily as a vehicle for achieving professional objectives.

For example, if a professional objective is to increase an employee's proficiency with Adobe Photoshop, I will look for a way to tie it to a work objective, either by finding an assignment that requires greater proficiency with Photoshop or by modifying a current assignment. Without a tie-in, it is very hard to keep the discipline necessary to make progress.

Whenever an employee says something like, "I'd like to learn more about XML," you can bet nothing will happen until he or she gets a job assignment that requires that skill. I do not stop employees from having that kind of objective, but when they do, I try to find a practical application to exercise that skill.

Another good way to get the ball rolling on any objective is to have the employee identify a concrete next step. Even a step as simple as "sign up for Perl 101" gives the employee a specific thing to do; without that next step, the probability that anything will happen goes way down.

Discussing rankings and ratings

If your company gives everyone a ranking or rating, you probably should lead with it. There is no doubt that nearly everyone is waiting for that piece of information, so I do not recommend withholding it. The downside, of course, is that if it is not what he or she is expecting, the entire discussion may be dominated by that one topic. If you suspect that may be the case, be prepared with a good explanation of why the rating/ranking is the way it is, and then use that explanation to direct the discussion towards the themes you want to discuss.

I try to preempt individual discussion about the ranking and rating process by briefing the team in a group meeting. Since nearly every description of a ranking and rating system I have ever seen reads like an example of bad writing, describing one to a group of technical writers with a straight face can be an adventure. Trying to do that one-on-one during a PE discussion wastes way too much time and gives ranking and rating a weight it does not deserve.

Another plus to having a general discussion in advance is that it is easier to set expectations at a group meeting than one-on-one. If everyone knows that only 10% of employees will be ranked in the highest category, you can set an expectation that even very competent people will be ranked in lower categories. Also, while I consider it bad form to blame the ranking system in an employee discussion, I am less scrupulous about acknowledging the system's shortcomings – and I have never met one that did not have shortcomings – in a group meeting.

Handling difficult situations

If you need to deliver a difficult message, for example a less than desirable rating, here are some guidelines:

- ► **Prepare:** Know what you need to say and how you are going to say it. Think about the questions you are likely to get and be prepared with answers. You cannot anticipate everything, but you can be ready for the most likely responses. If you are not sure how to deliver a strong message, talk with your manager and with HR. They should be able to give you some guidelines. If you are nervous about delivering a strong message, being prepared helps.

- **Be clear and to the point:** This is not the time to be vague. Most people are astute enough to see bad news coming, so there is no point in dragging it out with a long explanation. Here is a bad example,

> Jimmy, I know you tried really hard this year and there were some problems with the specifications from the engineering team over which you had little control, but, we felt that while those problems were a contributing factor in the lateness of the User's Guide, there were still factors over which you had control, but since you did not raise those issues in a timely fashion, it was not possible to fully factor in the delays, which meant that the project slipped, which is why your ranking fell this year.

I am embarrassed to say that at times I have done as badly or worse, usually in inverse proportion to the amount of preparation. Here is a better example:

> Jimmy, your ranking fell this year because you did not meet your schedule for the User's Guide.

You will have plenty of time to get into the inevitable discussion about contributing factors, etc., but a clear statement gets the issues out on the table.

- **The news is not a negotiation:** What you do about the news may be negotiable, but the facts are not.

- **Do not deflect responsibility:** You may have spent hours losing an argument with the rest of management team over this person's rating, and you may disagree strongly with the result, but you need to take responsibility for it anyway. If you believe there was something improper about the result or the process, talk with your manager or HR, otherwise, you need to own the result and present it as yours.

► **Be specific about what needs to happen:** This is especially important for the extreme cases. If someone's job performance is substandard, that person needs to know it and needs to know what specific actions are necessary to get his or her job performance back to an acceptable level.

► **Be specific about consequences:** If specific actions are required to save someone's job, then that person needs to know both the actions required and the consequences for inaction.

► **Be positive:** This can be tough, especially when the employee is on the brink of being terminated. It is very easy to fall into the mindset that you are just going through the motions because the company makes you go through a bunch of steps before you can terminate someone. But, besides covering the corporate derriere, these steps give the employee an opportunity to redeem him or herself. People do change, and it is only fair to be as positive as possible about their ability to change.

You may find yourself in a difficult situation that you did not expect. I can remember several discussions where I thought I was delivering a positive message and found myself faced with an employee who thought I was delivering a negative message. For example, just because you worked hard to get Eddie into the second highest rating category does not mean that he will be happy about not being in the highest category.

When that happens, I simply listen to what he or she has to say, then react honestly. That can be tough, especially if Eddie did not get rated in the highest category because you were not a good enough negotiator in the ranking meeting or your manager heard negative comments and gave them more weight than you did. In those cases, let the "Do not deflect responsibility" guideline lead you.

Do not expect an immediate resolution. If your explanation is accepted and you can see that the employee is okay with the result, fine. Otherwise, schedule a follow-up meeting so you can discuss things after he or she has cooled down and had time to think things over.

Wrapping up the discussion

Wrap up the discussion by reviewing the actions each of you will be taking. Think of this like any other meeting that generates action items.

Summarize them and make sure you both agree on who is taking what action. If there is disagreement over some point, most written PE forms provide a place for the employee to make comments. I encourage employees to write comments whether we agree or not, but it is especially important if the two of you disagree.

Finally, take advantage of the discussion to get to know your employees better. The PE discussion gives you a rare opportunity to talk about things that are not directly tied to project deliverables. Now is when you can get some idea of what an employee's longer term ambitions are and where he or she would like to be in a few years. When you know what an employee aspires to, you are in a better position to help with his or her professional development, and you are in a better position to plan the evolution of your team. While there is nothing stopping you from discussing these kinds of topics anytime, in practice that rarely happens, so take the opportunity.

Employee Ranking and Rating

So far, I have drawn a positive picture of the PE process, with only a few clouds. If approached seriously, the process described in the previous sections will give your employees a useful picture of their job performance and guidance for their professional development.

But, there is a dark side to the PE process. Between the time you complete the written PE, and the time you hold the PE discussion, the odds are that you will need to meet with your department management team – this includes you, your peer managers and your common manager at the next highest level – to assign ratings and rankings. This is where the PE process usually dives headfirst into the weeds.

Companies give lip service to an idyllic view of the PE. However, the real value of the PE process for the company is to sort employees into winners and losers. That is what managers are actually held accountable for, not the quality of the written PE or the quality of the discussion with the employee.

Sorting employees distorts the process. If you are an employee, you are unlikely to admit to any weakness that might push you down the list. If you are a manager, you need to communicate weaknesses to your employees, but you are also looking out for them as they get sorted in with employees from other groups. If you are forthright about your teams' weaknesses and other managers are not, your team as a whole will suffer.

What are ranking and rating?

Before getting into a discussion of how best to manage this distortion, let's look at the two common ways that companies sort employees, ranking and rating.

- ► **Ranking:** A list of employees in order of job performance from highest to lowest. For example, Freddy is ranked third in his team.

- ► **Rating:** A set of named buckets, high to low, assigned to employees based on their performance. For example, "Exceeds Objectives," "Meets All Objectives," "Meets Most Objectives," and the dreaded "Needs Improvement."

It is not unusual to combine these methods. After all, companies really want a ranking. It is very convenient for lay-offs and allocating goodies. But for an employee a true ranking, "you are number 44 out of 78, right between Betty and Veronica," is at best worthless and at worst strongly demotivating. Therefore, companies often create a ranking, then allocate ratings based on some formula (for example, 10% in the highest rating, 30% in the next highest, etc.).

Problems with ranking and rating

Some employees like the idea of a competitive workplace with relative rankings and ratings. They like to see how they are doing compared to others and they work hard to improve their position. In my experience, however, this group is in the minority, and a good portion of that minority only gets a short-term boost in performance.

In practice, the ranking and rating process is good for the top-rated people, since they get the goodies, and the bottom-rated people, since they get a strong message that they need to change. But, the top-rated people are nearly always already strongly motivated and do not need a ranking or rating to stay that way. And the bottom-rated people should be getting that strong message anyway, or you are not doing your job as a manager.

For everyone else, and that is most of your team, a ranking or rating is at best a no-op and at worst a demotivator. If they accept the bucket they have been placed in, they will listen to your feedback. If they do not accept their bucket, they probably will not hear a word you say.

Besides the demotivating aspects of ranking and rating, there are some other problems should be aware of:

- ► Ranking and rating are easy to confuse. Some companies will define a group of rating categories (for example, "Exceeds Objectives," "Meets All Objectives," etc.), then force a percentage distribution across those categories (for example, 10% "Exceeds Objectives," 30% "Meets All Objectives," etc.). As a result, you may find yourself trying to explain why an employee was "rated" "Meets Most Objectives" when he or she clearly met *all* objectives.

- ► Rating categories are often fuzzy. If your categories are: "Exceeds Objectives," "Meets All Objectives," "Meets Most Objectives,"and "Needs Improvement," where do you put an employee who exceeds expectations in two areas, meets expectations for all but one other area, and needs improvement in that last area?

- ► Ranking is difficult, if not impossible, across disciplines. For example, how do you rank order a technical writer and a firmware programmer working on two different products?

- ► Forcing a distribution across a set of ratings only works when the group is large enough to eliminate the distortions that arise in smaller groups. Depending on who you listen to, you may need 100 or more people to get a reasonably fair distribution. It is unlikely that any manager who has a department of 100+ people will know all of them well enough to arbitrate a fair distribution. Therefore, the distribution gets pushed down to smaller groups, where it is almost surely unfair.

- ► There is no completely fair way to merge ratings from sub-teams into a larger group. The standard way to do this is to force each sub-team to have the same distribution of ratings, then fight out the borderline cases. But, then you are back to the small group problem.

Overall, ranking and rating work against employee development. But, in many companies you have no choice but to work with them.

Formulating rankings and ratings

While the methods used to formulate rankings and ratings vary widely, there are common elements to most of them. I will focus on those common elements and provide some suggestions that should help no matter how your company handles the details.

Common elements

Though there are innumerable variations, most ranking and rating systems I have seen have a set of common elements:

- ► A rating system that puts everyone into one of several categories. The categories may be descriptive or non-descriptive, but are always ordered.

- ► A recommended or forced distribution across the rating categories.

- ► A process for assigning ratings to employees.

- ► Reviews by HR and higher level management to ensure that the rules have been followed.

- ► A formula for relating ratings to rewards – salary, bonuses, stock options, and so forth.

- ► A process for communicating the results to employees, usually through the PE discussion.

Ranking and rating meetings

The core of the ranking and rating process is typically a *battle royale* fought out in a management team meeting. Here is where the fate of your team is determined. If you succeed, your team will get more goodies and presumably be happier; if you fail, you will have an unhappy crew.

While I have heard gruesome stories of shouting matches, betrayal and back-stabbing, they are the exceptions. Managers usually understand that they need to work with their peer managers every day, that not everyone on their team can be rated in the top category, and that not everyone on the other teams is an idiot. I have also found that the manager in charge of the meeting nearly always steps up to the responsibility of averting fistfights and hair-pulling.

Here are a few guidelines to help you get the most out of these meetings:

- **Understand the deliverables:** Make absolutely sure you understand what your management and HR expect you to deliver. Will you be expected to deliver a set of ratings, or will you be expected to deliver a list of people in rank order? As you might guess, the former is much easier than the latter.

- **Understand the process:** Make sure you understand how the meeting will be conducted. If the process is vague, you can be sure there will be problems. If you are in control of the process, tighten it up; if not, work with the manager who is in control to encourage a tight process.

- **Be clear about who is the final arbiter:** There will be disputes that need to be resolved. The highest level manager in the room needs to take the responsibility of being the decider. If that is not made clear in the process, you can bet there will be a problem somewhere along the line.

- **Meet on an equal footing:** If you have a distributed management team and cannot arrange a face to face meeting, use a teleconference with everyone on the phone. If you try to bring in one or two managers by phone while the rest of the team is together in a conference room, the managers who dial in will be at a disadvantage.

- **Remember you are a team:** Though I have spent most of this book talking about the team of people who work for you, you are also a member of a team of managers reporting to a higher level manager. No individual's rating is important enough to risk destroying your relationship with your peers.

Some final thoughts on rating and ranking

When I described my views on rating and ranking to a former colleague, he took me to task for not recognizing that rating and ranking are useful as a stimulus to improve performance. In his view, knowing where you are ranked can cause you to improve your performance.

He has a point, and I do not deny that there are people who motivate themselves to improve their ranking or rating. But, I think the majority of people find their long term motivation in other places.

As a manager, I think the best time to focus on rating and ranking is when it reinforces a point that is not getting through by other means. For example, if you have an employee who is not meeting a reasonable standard of performance, pointing out that his or her job performance is at the bottom of the department can be a powerful stimulus for change. Conversely, if you have an employee who excels, but is over-critical of his or her performance, revealing a high ranking may give him or her a boost.

Like any other tool, ranking and rating can be used or abused. But, recognize that in a lot of ways they are the chain saw of management tools. They are great for some tasks, but if you do not know what you are doing and do not use them with great care, someone is going to get hurt, probably you.

Managing Projects

9

Development
Methodologies

Any development methodology is better than none.
— author unknown

I spent a great deal of time early in my career studying development methodologies; I read books, attended conferences, wrote papers, and in general steeped myself in the field. While the urge has abated somewhat, I still keep up with trends. But, the epigraph here is not far from the truth. The benefit from any methodology comes from the discipline of figuring out what you are going to do and how you are going to do it. The details of how that methodology gets you there, while important, are secondary.

Development methodologies come in two basic models: "Sequential" and "Iterative." The sequential model is characterized by a sequence of phases, including, Requirements, Design, Implementation, Test, Deployment, and Maintenance. In a pure sequential model, each phase is completed before the next one begins. The iterative model is characterized by short cycles, typically two to four weeks, each of which includes Requirements, Design, Implementation, Test, and Deployment of a subset of features.

There are modifications and variations galore. Most methodologies that follow a sequential model provide for development iterations with unit testing to measure progress. And, iterative methodologies start with at least a high level idea of what the result will be, even if they do not start with detailed requirements.

Sequential Model

The sequential model grew out of traditional, and time-tested, engineering methodologies. If you are building a bridge, you will be in trouble if you do not have detailed engineering plans before you start pouring concrete. Therefore, even for software, which is considerably more malleable, the sequential model begins with a thorough requirements analysis and detailed design before development begins.

While there are myriad variations, the sequential model will generally have the following phases:

- ► **Requirements Analysis:** An analysis of requirements, which are then documented in a requirements specification. In this phase, the documentation team identifies the audience for the product or service and determines the type of information that audience will need.

- ► **Design:** A detailed description of the system design. For larger systems, the design process includes detailed designs for each subsystem. In this phase, the documentation team designs its deliverables.

- ► **Implementation:** The actual development of the product. For software projects this includes software programming and usually unit tests (a unit test is a test of a single feature or part of a feature). In this phase, the documentation team develops content.

- ► **Test:** Testing of the various parts of the project, both apart and separately. Most methodologies test separate units as part of implementation, then test them as they are integrated into the whole. Therefore, this step is sometimes called "system" testing to differentiate it from unit testing. In this phase, the documentation team holds content reviews, tests content with the product, and prepares deliverables for publication.

- **Deployment:** Delivery of the product to customers. In this phase, the documentation team may deliver physical media or electronic content, like a help system, with the product, or they may deliver content separately, for example on a web site.

- **Maintenance:** Problem solving, bug fixing, support, returns, and so forth. In this phase, the documentation team tracks changes and updates content as needed.

If your project is small, or if the constraints are well understood, the sequential model can work well. It has the advantage of providing a view of the entire project relatively early, and it is reasonably intuitive for most managers. In my experience, higher level managers like this model because they can see a plan for the entire project up front.

The disadvantages stem from a lack of flexibility. If requirements change, or more likely, if requirements become better understood over time, it can be difficult if not impossible to adjust. And, seeing the entire project from the start may give you the illusion that things are under control, when in fact they rarely are.

Iterative Model

The iterative model, which is exemplified by "Agile" methodologies like *Scrum* or *Extreme Programming*, divides a larger project into short (two to four week) cycles or "sprints". Each cycle results in a potentially deliverable product. Unlike sequential methodologies, iterative methodologies go through analysis, planning, design, and implementation phases for each cycle. They evaluate progress, often with the customer, at the end of each iteration.

Iterative methodologies address the flaws of the sequential model by explicitly providing for changes in requirements and design as the project progresses. However, it is not all a bed of roses. As with most new methodologies, there is the good, the bad, and the ugly. Here are some things to consider:

The Good

Most Agile methodologies integrate documentation development tightly into the process. For example, Scrum puts all work items, including documentation deliverables, technical reviews, and even time spent meeting with subject matter experts, onto what they call the "backlog." The backlog defines all of the work to be performed in a given cycle. A cycle is not considered complete until the backlog is exhausted.

Communication is typically more open and frequent. Scrum has daily meetings, called "scrums" after rugby scrums. These meetings, which last only 15 minutes, give every member of the team the opportunity to describe what happened yesterday, what is planned for today, and what work is blocked. This helps ensure that typical documentation problems, like slow reviews or lack of access to subject matter experts, are addressed and resolved quickly.

Documentation of plans is lighter, more flexible, and more frequently updated, and the plans themselves are smaller in scope. This should lead to more accurate scheduling, though it is far from a silver bullet.

The Bad

Re-work in the engineering is much more likely to result in re-work for the documentation. In a traditional methodology, requirements have time to settle down before development, and development has time to settle down before documentation. With Agile methodologies, the overlap of requirements, design, development, and documentation means that false starts and re-work can ripple through the entire chain.

The Ugly

Some teams seem to think that Agile means you do not need plan. You can run into everything from a mild disdain for long range planning to total chaos. In a good project, you will be able to plan deliverables at a high level, but you may not be able to design their structure until late in the game.

A modular documentation structure will give you more flexibility in an Agile environment, particularly if you are delivering help or web pages. Modular methodologies allow you to develop bottom up and build your deliverables later in the process.

The real world

I fall somewhere in between when it comes to project methodologies. I like the free wheeling style and tight interaction of the iterative methodologies, but I also like to have a clear vision of where the project is headed.

In practice, I try to do as much requirements planning as possible up-front, then implement using an iterative methodology where those requirements could be adjusted as needed along the way. This matches my overall philosophy of management, which is that you need to know where you are going at all times, yet remain open to adjustments in both your destination and your path to that destination.

In the real world, you will probably have little input into the development methodology chosen by your project managers; you will need to fit into an existing structure. However, project managers usually care more about the reports and other artifacts than the details of how you get things done. If you provide the right reports and deliver as promised, you often have a lot of flexibility in how you run your team.

Therefore, a modified iterative methodology that gets a plan in writing up front, but treats that plan as adjustable over time, is likely to fit with everything except an extreme sequential or extreme iterative methodology.

Chapter 10 describes a method for planning that will work for most sequential or iterative methodologies. You may need to adjust some of your documentation, but the basic principles should apply across any but the most extreme versions of either model.

10

Project Planning

In preparing for battle I have always found that plans are useless, but planning is indispensable.
— Dwight D. Eisenhower

Eisenhower put it bluntly, but accurately. Planning is essential. There is no other way for you to allocate your resources, estimate effort, or otherwise manage your team. You need to know where you are starting from, where you are going to, how you are going to get there, and when you are going to get there.

The danger comes when the plan is documented. Consider the following: The project manager[1] tells you to create a documentation plan. You follow instructions and write a highly detailed plan, which gets skimmed, filed, and forgotten. Months later, when there is a problem with the project, the project manager pulls out your plan and uses it to convict you of failing to meet your commitments.

Even if this has not happened to you, yet, you have probably seen it happen to someone else. As a result, many managers try to duck the written plan altogether or make it as content-free as possible. Despite

---

[1]I will use the terms project manager and project management to refer to the people who manage the project for which you are creating content.

the potential downsides, there are good reasons for documenting your plan:

- ▶ Your plan's first job is to identify your objective. To quote Lewis Carroll, "If you don't know where you are going, any road will take you there." You must know where you are going; planning helps you define your destination, and your written plan communicates your journey to the rest of the organization.

- ▶ Unless you have a very small group, you will have too many projects to keep in your head all the time. A well-written plan keeps the information you need close at hand.

- ▶ Every plan makes assumptions, weighs risks, and considers contingencies. Your written plan communicates these caveats to management and gives you at least some cover when things go awry.

- ▶ Your plan helps you manage expectations. You can be sure that if you do not manage expectations, they will grow without bound.

- ▶ Your plan provides ammunition when things change or when you get handed an arbitrary schedule from above.

- ▶ In any organization that tries to plan, or even gives lip service to planning (and these days that is most organizations), you are going to need to write a plan. If that plan tracks changes in schedule, products, and deliverables, you will have a better shot at handling those changes.

So, you need to both plan and document your plan. This chapter will discuss both and help you make sure that your written plan works for you instead of against you.

Rules of Thumb

Before we look at the specific parts of a plan, let's consider some rules of thumb:

- ▶ **Write the plan for yourself:** Make sure you answer the questions *you* need answered; the chances are that when you do that, you will also answer the questions that project management wants answered.

- ▶ **Keep it up to date:** The plan is not finished until the project is finished. If you do not keep your plan alive and up to date, you will regret it when project management hits you over the head with your own, now obsolete, words.

- ▶ **Keep it simple:** Since you are going to update the plan frequently, you do not want a plan that takes more time to update than to execute. You can use a product like Microsoft Project, but often a spreadsheet or text document with the right information will do the job.

- ▶ **Keep it public:** Make sure the latest version of your plan is available to anyone who needs it. Keep it on line and easy to find. I like the way standards organizations solve this problem. They create a URL where you will always find the latest version and place that URL prominently at the beginning of the document. They may keep older versions available at other URLs, but the most recent version can always be found in the same place.

- ▶ **Cover your posterior:** Document as many assumptions, prerequisites, risks, and contingencies as you can. Documenting assumptions and prerequisites tells the project what you need to complete your responsibilities, and gives you leverage when things do not go as planned. Documenting risks and contingencies helps you be prepared when things go wrong.

Defining Objectives

A good plan is like a good newspaper article. It defines what is happening, when it is happening, who is doing it, how they are doing it, and where they are doing it. And, like a newspaper article, a good plan leads with the most important information: your objectives and the deliverables you will create to meet them.

Defining your objectives lets everyone know what you are doing, and just as important, what you are not doing. Often your biggest surprises occur when someone is convinced you are going to do A when you really plan to do B. Being clear about what you are, and are not, doing will help avoid this kind of surprise.

Defining objectives and deliverables should be a collaboration with the development team you are supporting. Bringing them in early helps guarantee buy-in and sets expectations. You should set expectations for what you will deliver and also for what you expect the development team to do. The latter includes their participation in reviews, testing, consultation, and so forth.

While you need to collaborate, you also need to drive the process; do not leave it in the hands of product developers or their managers. They will either tell you that their project deserves a 300-page book, a five-day course, online help, and a dedicated team of writers, or that the product is so easy to use that it does not need documentation at all. While you need to listen to your development team, keep a firm hand on expectations.

Objectives frame your work. They do not need to be detailed or lengthy; in fact, they must be simple and clear. You and your team should be able to explain the objectives in an "elevator statement," that is, it should not take more than a 30-second elevator ride to tell someone your objectives for a particular project.

Here is a sample objective statement:

> Update the ABC product user guide for version 1.3. Updates include adding the X and Y features, fixing currently open severity 3 and higher documentation bugs, and updating content to match the new style guide.

Though concise, this objective clearly identifies which deliverables you will update and how you will update them. It also defines what you will not do; you will not fix lower severity bugs, and if there is a feature "Z," you will not include it.

Defining Deliverables

I like to define deliverables using an annotated table of contents (TOC). A good annotated TOC shows what you are covering, the level of detail, the size, and for updates, what is being added, changed, or staying the same. Example 10.1 shows one way of structuring a TOC for a hardware product.

Example 10.1. Sample Annotated Table of Contents

FooBar 8250 Installation Guide, 5th Edition

1. **Preface.**

   a. **Purpose and Audience:** [*2 pages*] – Unchanged from last edition.

   b. **Typographical Conventions:** [*2 pages*] – Use standard library.

   c. **Changes in 5th Edition:** [*4 pages*] – New; replace previous version.

2. **Site Preparation:** [*12 pages*] – Minor changes from previous edition.

   Environmental requirements for the FooBar 8250, including power, cooling, and weight-bearing requirements.

3. **Unloading the Hardware:** [*11 pages*] – New; replace previous version.

   Unloading procedures for the new Ultra-2281 packaging system.

4. **Installing Software Updates:** [*23 pages*] – New section.

   Install software updates using the network connector.

This example includes page counts, plus a summary of what will be done with each section. For new sections, you can use the page counts directly in your estimates; for updated sections, you will need to estimate the level of effort required. Depending on how you define "minor," you have between 40 and 50 pages of work to do on this deliverable. In this case, I would probably start with an estimate of 50 pages of work and only reduce the amount if I became convinced that the "minor" changes were indeed minor.

The annotated TOC works even if your methodology is not book-oriented. Regardless of how you develop content, it will eventually be delivered to customers in some form, be it web page, help system, wiki, print, or whatever. Those deliverables are what project management will want to see in your plan and schedule.

Creating Schedules

The schedule is the one part of your plan that project managers will always examine closely. This section looks at the dimensions of scheduling, estimating effort, and identifying dependencies.

Managing the dimensions of scheduling

Scheduling works in three dimensions:

- ► **Content:** The scope and quality of your deliverables.

- ► **Time:** The amount of time available.

- ► **Resources:** The people who will develop the content, plus the cost of tools, space, services, and other expenses.

You never have equal control over all three dimensions. There is a saying among project managers: "You can have it good, cheap, or fast; pick any two." While that is an oversimplification, any project has to balance the time and resources available with the quantity and quality of the deliverables.

Because of this, many methodologies include a *Flexibility Matrix*, which documents the degree of flexibility you have in each dimension. Table 10.1 is typical.

Table 10.1. Flexibility Matrix

Dimension	Least Flexible	More Flexible	Most Flexible
Content			✓
Time	✓		
Resources		✓	

In theory, a flexibility matrix can help you make trade-offs. In practice, for documentation managers the flexibility matrix nearly always looks like this example. You will nearly always have the most control over content, less control over resources, and barely any control over time.[2]

---

[2]While this reality can be discouraging, understanding the three dimensions of scheduling has benefits, as we will see later.

When it comes to documentation, most project managers reduce "good, fast, or cheap," to "fast and cheap." While an important part of your job is to convince project managers that this is not always true, most of them will start from that point of view. In fact, the chances are excellent that most of the time you will be presented with the project schedule as a fait accompli and be expected to write a documentation plan that fits.

Clearly, shoehorning your plan into a pre-determined schedule will not always work. If you cannot deliver an acceptable set of deliverables with the available resources within the mandated schedule, you need to take action. I will return to this question in the section titled "Dealing with Unreasonable Schedules" (p. 114), but first let's look at the elements of building a schedule: defining deliverables, estimating effort, identifying assumptions, risks and contingencies, assigning resources, and combining schedules.

Scheduling deliverables

Scheduling starts by breaking down your deliverables as far as possible; the more granular your estimate is, the more accurate it will be. As discussed in the section titled "Defining Deliverables" (p. 104), I like to break deliverables down to a point where I can get a reasonably accurately page count. A page is not necessarily a printed page – it could be a help screen, web page, or a modular block of content – but it is an amount of content that would take about one page if printed.

What matters most here is to use a consistent measure that is not too large. Since writers understand pages, they often find it easy to use pages as a yardstick, even when the deliverables do not explicitly have pages. That said, some other measure can work as well, provided it is consistent and not too large.

Joel Spolsky,[3] recommends breaking work down into parts that will take no more than a few hours to complete. While this may sound excessive, it really is not. If you have an accurate page count and use the "two pages per day" rule of thumb discussed below, the level of granu-

---

[3]Spolsky's website, Joel on Software [http://joelonsoftware.com], and book [34] focus on software development, but should resonate with any manager.

larity will be about four hours, which is consistent with Spolsky's recom-mendation.

Estimating effort

The only rule of thumb I have ever used successfully for estimating effort is "two pages per day plus a fudge factor." This means researching, writing, editing, and delivering those two pages, not just writing them.

Of course, that fudge factor is critical, and unfortunately, it is the hardest thing to estimate. It depends on the individual writer, the quality of the product requirements, the probability the requirements will change, the probability development schedules will change, and the way your organization manages schedules.

Start with the product requirements. Look at how detailed the require-ments and schedule are, how sane the overall project looks, and how well it covers contingencies.

Next, consider how likely the requirements are to change. If you have been in the organization for a while, you will know. If not, ask around. It will not take long to figure out how stable requirements are in your organization.

A Warning on Metrics

We have a dilemma; how do you keep his-tory, but avoid meas-urement dysfunc-tion? The best you can do is delegate estimation to writers, then evaluate the ac-curacy of estimates and not "productiv-ity."

Then, look at how your team and the writer assigned have handled past projects. Keep track of how closely past estimates have matched reality, then use that information to improve future estimates. One way of doing this is called: "Evidence Based Scheduling,"[4] which tracks and compares each participant's initial estimates with his or her actual results. Over time, you build a history that lets you adjust estimates based on past performance.

Finally, take a step back and look at how project management handles schedules. In

---

[4]Joel Spolsky has an excellent description of this method at http://www.joelonsoft-ware.com/items/2007/10/26.html

my experience, project managers treat (or mistreat) schedules in one of three ways:

1. The schedule is inviolable no matter how crazy it seems. Management will flog developers, compromise quality, pare features, and do anything else necessary to get the project out on time.

2. The schedule is inviolable no matter how crazy it seems, until management concedes, usually late in the game, that for "unforeseen reasons," the original schedule must slip. This category has a couple of sub-categories:

    a. The schedule slips and/or features are pared in repeating cycles until the product is releasable. If you are lucky, each cycle is smaller than the last. In my experience, most development teams work like this.

    b. The revised schedule, either immediately or after a couple of iterations, is re-set in stone and management reverts to the first category (developer floggings).

3. The schedule is developed sanely, updated frequently, respected by the whole team, and when, rarely, it slips, the organization handles things maturely, making sensible feature/schedule trade-offs and maintaining quality.

If your organization falls into the third category, consider yourself *very* lucky and do your best to meet the same standards.

The chances are your organization will not be in that category. To get the real story, skip the managers and talk with product developers, testers, and other writers. Most managers, especially those in the first category, believe they are in the third category. The most blatant tip-off is when they proudly tell you they have *never* missed a deadline.

No matter what you do, it will take a few projects with your team before you start getting estimates that you can feel good about. There are too many variables to apply a set formula and expect good results. Just do your best and refine your estimates over time.

What is important is to get as good a handle as you can on the magnitude of the job. That is why I use the two-pages rule. It is simple, clean, not too bogus, and it seems to work pretty well. Most importantly, if you can count the number of pages, you are getting to a useful level of granularity.

Identifying dependencies

Dependencies are prerequisites that must be met before you can complete some part of your plan. They identify places where you may become blocked by factors outside your control. For example, if you cannot begin work on a documentation deliverable until the engineering requirements for the associated product are complete, you might include a statement like, "the start date for deliverable X depends on the engineering team delivering the completed requirements document for feature Y to the writing team by *date*."

I give dependencies a prominent place in planning and in the written plan. Along with contingency planning, they form your primary defense against factors you do not have direct control over. Take the time to capture as many as possible. It is not enough to simply document dependencies in your plan. Remember that no one will pay any attention to your plan until things go wrong. Therefore, you need to make sure that everyone responsible for a prerequisite or dependency acknowledges their obligation. Then, you need to track dependencies as carefully as you track your own team's milestones.

You may find yourself having to start without everything you need. If this happens, make sure your engineering contacts are aware of the situation, but try to be adaptable. Often you can get started on aspects of your work – possibly structure or boilerplate – that do not require a particular dependency, and it is not unusual to be able to accomplish much more than you expected without those dependencies. That said, when you have a hard dependency, make sure you document it and make sure the people you are depending on understand the dependency and are committed to satisfying it.

Assumptions, Risks, and Contingencies

Identifying and documenting assumptions, risks, and contingencies is a critical part of your planning. You need to get every spoken and unspoken assumption and risk on the table, along with contingency plans for dealing with them.

Assumptions

Some assumptions do not require an assessment of risk or a contingency plan. These are assumptions like, "we will use standard XYZ corporation processes." This kind of assumption acts as a shorthand for what might otherwise be a lengthy explanation of your processes.

The assumptions that matter are the ones related to prerequisites and dependencies; the ones that have an impact on your deliverables or your schedule and are outside of your direct control.

For example, if the writer responsible for a particular section will not be able to work on it until another project completes, state that as an assumption. And, when there is a reasonable probability that the assumption will prove false, you should document the risk and a contingency plan.

Risks

To document a risk, you need to consider three things: the likelihood of the event occurring, the impact if it does occur, and what you will do if the event occurs. For example, if you need a working prototype of a piece of hardware by a particular date to write a module, then there is a risk that the prototype will not be available. You might document this risk as follows:

Assumption:	Working prototype will be available by May 15.
Risk:	Prototype delayed or not fully functional.
Likelihood:	Low (under 10%)
Impact:	High. The documentation that depends on the prototype is in the critical path for the project. Any delay will slip the schedule unless additional resources are made available.
Contingency:	If the prototype is delayed less than two weeks, and the delay is announced at least two weeks in advance, we will bring in an additional writer, split the work, and maintain the original schedule. Cost for the additional writer will be $AA.
	If the delay is more two weeks, we will do X at a cost of $BBB per week. If the delay is announced later than two weeks in advance, we will do Y at a cost of $CCC.

Contingency plans

Contingency plans can be tricky. In the case above, you probably have several options for each case. For example, you might shuffle assignments, delay another project, or reduce content. The amount of detail you include should reflect the likelihood of this risk occurring and the impact if it does.

You should also highlight contingency plans for risks whose impact is disproportionally high. Project managers often assume that a slip of X days in a dependency will result in an equivalent slip in your schedule. If this will not be true, make sure you state this clearly.

On the other hand, there is no point in documenting a risk that has minimal impact, especially if the impact is internal to your team. There is also no point in taking too much time with a low likelihood risk, even if the impact might be high. For example, you should have a high-level contingency plan for the possibility that a member of your team leaves with no notice or is incapacitated, but I would not list that in the documentation plan as a risk.

The one exception that comes to mind is when you know someone will not be available after a particular date, for whatever reason, and therefore a slip in some dependency that pushes the schedule beyond that date will have a bigger impact than just the original slip.

Once you have identified and documented assumptions, risks, and contingency plans, you are not done. You need to make sure that the people and organizations you depend on acknowledge and accept their responsibility. And, you need to follow up regularly. Some organizations review risks and assumptions periodically. If your organization does this, make your assumptions and risks part of that review. If not, track them yourself.

Assigning Resources

Once you know what you need to do, you need to decide who is going to do it. Depending on your organization, that may simply mean assigning writers who are already part of your team. On the other hand, you may need to make a budget to bring in new hires or outside contractors.

I prefer to assign writers to projects and let them define the deliverables, create estimates, work out their own schedules, and write the document-ation plan. This reinforces ownership of the entire process and also puts the estimate in the hands of the people most likely to get it right. Of course, with a less established team or less experienced writers, you will need to be more involved, and if you are bringing in external con-tractors or new hires you may need to do all of the planning.

As a general rule of thumb, I like to make assignments as stable as possible and only change them when absolutely necessary. This may at times be a little less efficient on paper, since it is harder to drive out schedule gaps – those places where a writer has completed project A, but cannot start project B until some prerequisite has been satisfied. But I think it is more efficient in practice because you have continuity of projects and writers, which makes the development process more efficient. And, those gaps can be used to improve processes, take train-ing, or pay down documentation debt (see the section titled "Document-ation debt" (p. 114)).

Beyond the jigsaw puzzle of assigning work to writers, I also like to have people assigned as backups for each project. The idea is to have people who can pitch in and help if a project gets out of hand or if the person assigned gets sick or is otherwise unavailable for a while. One way of doing this is to group a set of related deliverables and assign it to a team of writers. While each writer has primary responsibility for some sub-set of those deliverables, the team as a whole is responsible for the entire set, and therefore is responsible for load balancing to make sure the project gets done.

Combining Schedules

If you manage more than one or two projects, you will need to create a combined schedule for your team. A combined schedule gives you an overview of what each member of your team is doing at any given time. This is useful for those always unscheduled moments when your boss comes in and asks you what everyone is doing. It will also help you balance workloads.

You can use a product like Microsoft Project to do this, but unless you are tracking a lot of projects and a lot of writers, you can probably get away with a shared spreadsheet or even a text file. What is important is that you have one place where you can see at a glance each deliverable, the writer working on it, and its schedule.

If you manage a large enough group, it may be helpful to use project management software to combine and track your schedules. If you do this, take time to learn the system well, but do not try to get too fancy. I have found that most of the bells and whistles in project management systems are not all that useful. If your system can maintain schedules from multiple sources, merge them, and print PERT or GANTT charts, you have probably got most, if not all, of what you need. You may not even need that much; I have managed 10 to 12 concurrent projects using only the most basic functionality of Microsoft Project.

Dealing with Unreasonable Schedules

If you get through all this planning and find that you cannot deliver an acceptable set of deliverables with the available resources within the mandated schedule, you need to speak out strongly and quickly. However, make sure you have some alternatives to propose *before* you raise a ruckus.

Project managers will expect you to exhaust all of the possibilities within your control before you come to them asking for resources or time. They will also expect you to look into all three dimensions of the flexibility matrix, content, resources, and time. When you do come to project management, you need to bring a proposal that describes what you have already done and alternatives that address all three dimensions.

As we have seen, the flexibility matrix will almost always have the same order, with content always being the most flexible in the eyes of project management. Therefore, you can be sure the discussion will center on a trade-off between content (both quantity and quality) and resources (both staff and schedule). You will want more resources; project management will want the same resources, but will be willing to accept less (or lesser) content. To help manage this trade-off, we need to consider documentation debt.

Documentation debt

Documentation debt is a term adapted from technology debt. In both cases, you defer work to shorten your schedule. The debt you incur depends on the nature of the work you defer. If you defer essential work, the results will be disastrous; most likely resulting in either product failure or lots of lost customers.

If you defer necessary, but non-essential, work, the debt can still be large. For example, if you leave out what you think will be infrequently used procedures, you will receive support calls from those who need those procedures. If you picked the wrong procedures to leave out, the cost of those calls could be high.

If you defer unnecessary work, then the debt may be low or zero. The catch is that, while it may be easy to separate essential from necessary work, it can be difficult to separate necessary from unnecessary work. And, it can be difficult to gauge the amount of debt you will incur in any given situation.

In many ways, documentation debt is like financial debt. If you cannot afford to purchase a home with cash, you look for a mortgage. In the long run a mortgage will cost you more money, but it will also enable you to live in the home while you pay it off. With documentation debt, you are deferring necessary work in order to get your product out the door, hoping that the benefit of getting the product out earlier will be higher than the debt incurred.

However, the analogy breaks down because financial debt can be quantified and is generally predictable, while documentation debt is not. Taking on documentation debt is like taking on a sub-prime mortgage with the lowest possible down-payment, an interest only monthly payment, a variable interest rate, *plus* a balloon payment that can be called in at any time. As we have seen from the mortgage crisis in 2008, the first three of these factors can be extremely dangerous. Adding a balloon payment (that unexpected situation where everyone decides they need some undocumented procedure you thought was unnecessary) makes it lethal.

This does not mean you should never incur documentation debt, nor that you will be able to avoid it in some situations. It does mean that you need to consider the likely consequences of leaving out content or omitting a quality assurance step. While it can be tough to be 100% sure you are making the right choice, it is also true that you can almost always trim or eliminate something. And, the amount of effort you give to editing, indexing, reviewing, and testing can vary and still result in acceptable quality.

At this same time, there are levels of documentation debt that it would be irresponsible to accept. You need to point this out to project management and do all you can to avoid incurring debt in those situations.

If you are forced to incur documentation debt, document the direct cost of the debt (that is, the cost of the deferred work) in your plan and try to get it into a future budget. You may also want to document as a risk the possibility that the debt will cause unanticipated expenses.

Working with project management

Once you have identified alternatives, including documentation debt for deferred content, it is time to work with project management to find a solution.

Preparation is key. If you go to project management with a recommendation that includes reasonable reductions in content that incur a manageable documentation debt, along with accurate estimates of what additional time or resources are needed, you will have a much better chance of getting what you need. If you just waltz in with a demand for more time or resources and no evidence that you have considered all three dimensions, you will probably get slapped around and told to "make it work."

Preparation is key. Yes, I am repeating myself. Every time I have gone to project management with a clear plan, backed up with reasonable analysis, we have been able to arrive at a solution that worked for everyone. Every time I have gone to project management unprepared, I have come back with a headache, and usually with an imposed solution that resulted in lower quality, tighter schedules, and overstressed, unhappy writers.

Miscellanea

In addition to the objectives, schedule, prerequisites, dependencies, assumptions, risks, and contingencies, here are a few miscellaneous things to consider:

► **Personnel:** Generally, I ask individual writers to write their own plans, so the personnel may be implicit. However, even when it is not, I do not name the writers who will be working on a project unless the project managers insist, which they sometimes do. If the specific writer has not been hired, or the assignment is not clear, I will often identify the skills required to do the job.

► **Budget:** If you are writing this plan to get funding for a particular project, then you will need a budget. You may also need it if you want to document the limits of what you will be able to do on the project. If you do not call out the budget separately, you should still state as an assumption the resources that will be applied to the project. For example, you might say that you will be devoting X staff months of effort across the Y months of the project.

► **Review and Test:** Make sure that documentation reviews and tests are included in your schedule and that participation from the architects and the development teams is included as a dependency. Since these activities can be scheduled, you should work with engineering to get them into their plans, too. It is all too common to get towards the end of a project and discover that no one has the time to do a technical review. Getting reviews onto the engineering schedule may not prevent this from happening, but it will give the reviews visibility and increase the chances that they will occur.

► **Approvals:** As part of gaining buy-in for your plan, you should make sure you get formal approval from all relevant parties, including those who depend on you for a delivery and those who you depend on in any way. The key is to make sure approvals and commitments are documented clearly and unambiguously, and are publicly acknowledged.

► **Boilerplate:** Most companies have at least a few required sections in plans that have no real reason for existence, other than a long-forgotten mandate from a long-gone manager. It is best to just accept the inevitable and waste as little time as possible.

Writing the Plan

I suggest dividing your plan into two distinct parts:

- **High level plan:** The high level plan defines the objectives, deliverables (at a high level), assumptions, schedule, dependencies, risks, and contingencies. This is the plan you share with the rest of the organization.

- **Deliverable design:** The deliverable design is the detailed design for each deliverable. This could follow the annotated TOC style described above or it could parallel your development structure. Your deliverable design will probably comprise several documents, one for each deliverable or logical group of deliverables.

The reason for dividing your planning documents is to keep the high level plan simple, short, and easy to modify. That does not mean you should hide your deliverable design; it is important and should be easy to find. However, project managers mostly want to see the schedule, and you want them to see the assumptions, dependencies, risks, and contingencies. If you add in the detailed design of each deliverable, your plan will get large and unwieldy, and the things you want project managers to see may get lost in the shuffle.

I have already discussed documenting your deliverable design (see Example 10.1, "Sample Annotated Table of Contents" (p. 105)). Appendix A, *Documentation Plan Template* (p. 239) contains a template for a high level documentation plan. There are many varieties of documentation plan that can give you a usable template. Consider this template to be one of many, and one that conforms to the philosophy described in this book.

Remember also that once you have written your plan, you are not finished. Keep it up to date and follow up on dependencies and assumptions with their owners. I will look into this in more detail in Chapter 11: *Tracking*.

11

Tracking

Even if you're on the right track,
you'll get run over if you just sit there.
— Will Rogers

Project tracking is your early warning system. All projects will go off the rails at some point or another; good tracking will keep you from going off in the wrong direction too often and will help you get back on track quickly when you do derail.

At its heart, tracking is simply keeping track of the *milestones* in your plan. If you have created a plan with frequent, easy-to-measure milestones, this is pretty simple. But, there is more to good tracking.

If you have ever tried to track a person or animal in the woods, it does not take long to see that there is much more information in a track than just a trail in some direction. With careful study and practice, you can learn how to determine what kind of creature you are tracking, how big it is, how fast it was going, which way it was going, and how long ago it made the tracks. You may also be able to tease out clues about its state of mind, for example, whether an animal has been injured or in a work environment a person is trying to conceal his or her path.

In the same way, as you track a project, you need to look beyond the simple yes/no questions and answers. Yes, they met the latest milestone, but the team has been working 80-hour weeks for the last month. Or, no, they missed the milestone, but it was because engineering slipped their schedule.

In the first case, you are probably in deep trouble; no team can keep up that pace for long. In the second case, you may or may not be in trouble. Sometimes an engineering slip will be a reprieve for your team and help you catch up; other times it will jam up work at the end of the project or conflict with time scheduled for other projects.

Basic Tracking

Basic tracking is simply keeping track of milestones and progress towards their completion. As a documentation manager, I have each writer track his or her own development milestones, and keep me informed on progress. I normally do not keep a record of milestones unless I have a writer who is inexperienced or unreliable. Even then, I have the writer keep track; I simply follow along more closely than I would with a more experienced writer. I do keep track of deliverable milestones and draft review milestones; not to second guess my team, but because other managers expect me to know this information.

The mechanism for tracking milestones is irrelevant, as long as it is visible to you when you need it. If you have many complex projects, you may want to use a product like Microsoft Project, which will give you all the features for schedule tracking you could possibly want, plus a bunch you have never thought of using. But unless you are building an aircraft carrier, it is probably overkill.

What is most important is to know what milestone any member of your team is working on at any given point, what the major issues are, and what the early warning signs for trouble are.

For example, let's consider a typical software project that needs a help system. Such a project will probably schedule a series of development cycles, each of which will add a set of new features, followed by several test and debug cycles. Your team will need to deliver new content for the help system at each cycle that supports the features that have been added at that cycle, plus updates and fixes related to previous cycles. You may also have additional content scheduled at various cycles.

Right up front there are a couple of things to keep in mind. When you are delivering content to a help system, you are tied directly into the software development cycle. You need to deliver content along with the software team on their schedule. That means that you will need to negotiate your deliveries and make sure your team meets them; otherwise, you risk delaying the software integration cycle.

Selecting milestones

Selecting which milestones to track is an acquired art. It is easy to identify the major milestones like completion of deliverables, but it is difficult to identify which of the many other potential milestones are worth tracking.

My rule of thumb is to track a milestone when another work item depends on it, when missing that milestone will trigger a contingency, or when the milestone is visible outside your team.

You also need to develop content in sync with the development team. If you are delivering a book or web page, you may be able to get away with scheduling your writing independently from the software development team. For a help system, you really want to have at least preliminary content in place for each feature as it appears in a development cycle. Therefore, as you negotiate schedules with the development team, you need to make sure your writers receive development changes in time to incorporate them into their deliveries. As you might guess, development teams will try to push their deliveries as late as they can; you can get caught short if you do not negotiate enough time up front.

In this example, the major milestones are easy to identify; they coincide with the development cycles. In between will be minor milestones that your team members will use to track progress leading up to their deliveries to each cycle. Normally, I will not personally track those minor milestones, unless I think there is a particular risk. Risks I care about include cycles that have too much content, writers who tend to over commit, and development teams that either underestimate the amount of change or neglect to tell you about changes.

Advanced Tracking

Effective managers go beyond simply tracking the work their team is doing and their direct dependencies. They also track up and across.

Tracking up

Tracking up means staying abreast of the larger project and its status. If your product is lagging in the market versus your competitors, you can bet that there will be significant pressure to tighten schedules. The same thing can happen if the trade rags begin to talk about your project. I am sure the pressure on the Microsoft Vista development team became excruciating as the press began to write stories about delays and problems.

To track up effectively, you need to pay attention to the parts of project team meetings that you normally would be tempted to sleep through; things like marketing, sales, manufacturing, and so forth. You also need to understand their plans. For example, if the marketing team plans to emphasize certain features, you may want to emphasize those features in your content. On the other hand, if they are de-emphasizing features or planning to remove them, the reverse may be in order.

This can be especially important as features or products become obsolete. Often companies make commitments about how they will notify customers about obsolescent features, then forget those commitments when it comes time to remove some feature. I have seen cases where the documentation team saved the day by highlighting a forgotten commitment. While it really should not be their responsibility to track obsolescence, documentation teams should be on top of what the requirements are for notifying customers and should not be shy about raising an alarm if the requirements are not met.

Tracking across

Tracking across means keeping up with the status of your peer teams, including product development, test, and integration. What you are looking for here are signs that they either are, or are not, meeting their milestones. This will help you know how likely it is they will meet your commitments, and it will help you when you meet with other managers. You will gain respect and leverage if you understand the pressures each of your peer managers is facing and do your best to avoid adding to those pressures.

While rare, it is not unheard of for a documentation team to help out a development or test team when they are under the gun and the documentation team has a little time. You can be sure that if you can help when they are in trouble, you will be in a much stronger position the next time you need something from them. And, the only way you will know they are in trouble is if you are tracking across.

Other considerations

At any given point, any project will have a set of issues that need to be worked. These are typically things like: delayed specifications, insufficient access to engineers, schedule shifts, and feature creep. Good project managers keep a list of these and work them aggressively.

As documentation manager, you need to make sure your issues are included in the project's issues list, and you need to work to resolve those issues; unless an issue that affects documentation also affects engineering, you are going to be the only person with any motivation to solve it.

If your project does not maintain an issues list, and you cannot convince them to do so, then maintain your own and review it with the project team regularly. In an agile environment, this probably means selecting one or two hot issues each day for your scrum. In a traditional environment where you have periodic status meetings, always review issues as part of your report. You should also follow up outside of formal meetings whenever you get the opportunity. When you start sounding like a broken record to yourself and the project team, you will know you are on the right track.

You should also involve your writers in tracking across and up. They should already be tracking their own work, but they can also provide extra sets of ears and eyes around the project.

Early Warning Signs

Early warning signs can be difficult to see. While issues and milestones are by definition visible (though you may need to work hard to keep them visible), early warning signs can be all over the map. When an engineering manager casually mentions in a meeting that an engineer in his or her group will be taking a family leave after delivering some code, you need to find out who will answer your writer's questions about that code after the engineer leaves. Or, if a project manager has trouble getting plans from one part of the development team, you can bet your writers will have trouble getting information from that team too.

Here are a few types of early warning signs to consider. However, this is far from a comprehensive list; you really need to keep your ears and eyes open.

- ► **Know who is reliable and who is not:** On any project team there will be people whose word is gold and others whose word is meaningless. You need to work with all of them, but a commitment from the latter should raise a red flag and cause you to take any preventive action you can think of. If you pay attention in project meetings or ask around about previous projects, you will know who is and who is not reliable.

- ► **Know your organization:** If you are new, this can be tough, but it is important to know how the project team manages schedules. Where do they fit in the categories described in the section titled "Estimating effort" (p. 108)? Will they pull out the whip and meet schedules even if they need to jettison features and quality? Or, will they slip the schedule? If they slip, is there a pattern? For example, do they always take one big slip, then hold firm, or do they allow a series of small slips? Do they slip early or late?

  Beyond giving you a strong clue as to how the organization will react when schedules are pressed, this will also raise a big red flag when project managers act contrary to the norm. For example, if an organization that slips early and often instead pulls out the whips early in the project, that may mean that higher level management is applying pressure. You may not be able to figure out why, but you can be pretty sure that you will want to do your best to stay out of the critical path for this particular project.

► **Know your schedule:** By this I do not mean just the milestones, though they are important. I mean, know where the schedule is ambitious and where it is slack. Know which milestones are hard and which are soft. Know where you can cut content if you get pressed. Once you know your schedule in this way, you have the flexibility to react to changes. And, you have information that can give you warnings. For example, if you have problems in a slack part of the schedule, it may be because you did not plan well and that when you hit an ambitious part of the schedule, things will go downhill fast. If that is the case, do not wait to re-plan.

► **Know your team:** In the previous example, the problem in the slack part of the schedule may be caused by a writer who is in over his or her head and needs help, it may be caused by engineers who are not as helpful or as available as they should be, or it may be caused by unplanned changes in functionality. Whatever the case, you need to heed the warning signs, investigate, and figure out how to get things back on track. Knowing your writers will help you diagnose the problem and get things back on track.

► **Know the scuttlebutt:** You can bet that your writers and their engineering colleagues know whether the schedule is realistic or not. They also know who is likely to be on time and who is always late. If you listen to your team, and have a relationship that encourages openness, you will have a much better chance of seeing warning signs in time to do something.

► **Know the vocabulary:** The first time I used the word "worried" as a manager, my manager reacted very strongly. To him, and to the organization, being worried was a strong statement that bad things were happening. If you were "concerned," you still had an issue, but it was much smaller. I needed to learn the right vocabulary to communicate successfully in that organization.

When managers talk, especially when they report status, they give clues both in their word selection and in their demeanor, and you need to notice both. Any organization will informally converge on meanings for various terms, usually in the same way my old organization differentiated worried and concerned. In addition, you need to pay attention to non-verbal communication. How confident do they sound? What does their body language tell you? What did they not say? All of these things are clues to what is going on.

Above all, pay attention. Listen to what your team tells you and encourage them to keep their eyes and ears open. Even though they can be deadly boring, go to project meetings and pay attention. This can be especially hard with phone meetings. Not only do phone meetings eliminate any possibility of picking up non-verbal clues, they also encourage half-hearted attention. Who has not listened to a phone meeting with half an ear while reading email or doing other work at the same time? While this seems like a good way to multi-task, you risk missing early warning signs.

12

Measurement and Metrics

What happens depends on our way of observing it or on the fact that we observe it.
— Werner Heisenberg

I worked as a software developer on an early version of network file sharing, the now common feature that lets you access a folder on another computer as though it were actually on your hard drive. During development, we met with Bill Joy, one of the founders of Sun Microsystems, to share information about the feature, which Sun was also working on.

Bill watched our demo with interest and asked one question, "How fast can you transfer data?" We had barely considered this question. We were consumed with the technical problem of making sure we never lost data and that programs worked the same way over the network as they did locally. Sun was carefully measuring speed, we were carefully measuring compatibility.

When our two products came out, Sun's product was without a doubt faster and ours was without a doubt more compatible.[1] Even though we were, on the surface, building the same capability, the result was significantly different because of what each company valued, and therefore, measured.

The Impact of Measurement

Measurements tell engineers what you value. If you keep track of the number of pages writers produce, they will produce lots of pages. If you keep track of how many errors get reported by customers, they will keep that number low. If you track every milestone, they will do whatever it takes to meet every milestone.

If you measure anything, the odds are overwhelming that that metric will improve over time. However, the odds are also overwhelming that something else will suffer. If you track how many pages get written, that number will go up, but readability will probably go down, since writers will not have an incentive to be concise. If you track errors, that number will go down, but the number of pages written per unit of time will also go down, as writers spend more time on each page looking for errors.

This will happen even if you do not use the metrics in performance evaluations. Writers understand that managers measure what they consider important. And they understand that metrics influence performance evaluations, even when managers claim they will not.

Managing technical writers by measuring output is a bit like controlling the shape of a balloon. If you press one spot, another will pop out. If you measure page count, writers will feel pressure to sacrifice other qualities of the content to increase page count. If you see that, and control some other measure, yet another dimension will pop out.

Management Strategies

Robert D. Austin's *Measuring and Managing Performance in Organizations*[3] formally makes the case that I have outlined above. That is,

---

[1] For those who care, our product was Remote File System (RFS) and the Sun product was Network File Sharing (NFS).

unless you measure every *critical dimension*[2] of effort, any measurement will result in dysfunction.

If you can measure every critical dimension – unlikely, though maybe just possible in a very simple job classification – then Austin would advocate doing exactly that; he refers to this as *Full Supervision.* In this case, every dimension that matters is measured and the manager has complete control. In addition to finding and measuring every critical dimension of effort, the manager also needs to set the right value for each of these objectives to optimize results.

If you do not measure every critical dimension, but try to manage as though you do, then you are exercising *Partial Supervision.* According to Austin, partial supervision will *always* lead to dysfunction because whenever you leave a critical dimension unmeasured, that dimension will be neglected in comparison with the measured dimensions, and because that dimension is critical, customer value will, at least in the long run, suffer. Many managers believe they are exercising full supervision when they are really exercising partial supervision, which exacerbates the dysfunction.

There may be workplaces where you can measure every critical dimension, but technical communication is not one of them. There are too many critical dimensions and many of them are too expensive or too difficult to measure or control. So, if full supervision is impossible and partial supervision is dysfunctional, what's a manager to do?

Austin defines one other mode, *Delegatory Management,* which eschews measurement in favor of delegating power to workers. It relies on them to select the amount of effort they will devote to each of the critical dimensions. It depends heavily on the manager establishing an environment of trust and mutual respect, and it also depends on internal motivation as a driving force. As we will see, delegatory management does not completely discard measurement; instead, it uses appropriate measures to provide information about the process and progress. But, it does not use measures to reward or punish individuals or groups.

---

[2] A critical dimension is one that if neglected will result in the customer receiving no value from the product or service.

Measurement Strategies

About this point you are probably thinking, "but, if I don't measure anything, how do I know how my team is doing?" Clearly some measurements are essential to tracking progress and cannot be ignored. The questions are, what do you measure, who makes the measurements, and how do you use them?

What to measure

I directly measure as little as possible. In fact, I have managed teams where I measured nothing. Instead, I delegated all tasks to team members, including all metrics. However, even if you delegate everything, there are a few things that must be tracked, even if only by an individual for his or her own purposes.

- ► **Milestones:** You have no real choice but to track milestones. As we have seen, you need to know as early as possible whether there is a problem you must address. However, I almost never personally track fine-grained interim milestones, and I frequently delegate even high level project milestones to the responsible engineer. Depending on the engineer's level of skill and experience, I delegate as much as possible, and stay in the loop only to the extent needed to report progress up the line.

- ► **Customer satisfaction:** While you may not want to use surveys or other formal means for measuring customer satisfaction, you do need to be responsive to customer concerns. I like to direct customer feedback to the person who is responsible for the content in question and let him or her handle the concern. While I may comment on how well someone handles a customer concern, I do not use the number of complaints or any other numerical measure in an evaluation.

In these two cases, and others like them, I am more interested in how writers handle the situation than I am in any kind of numerical measure. Milestones are met or missed for all sorts of reasons, not all them under the control of the writer. Customer concerns are equally indeterminate, at least with respect to number. There is no way you can learn much about writers by counting complaints, though you can learn something by seeing how they handle the complaints they get.

Generally, I avoid any measurement that has a numerical component. This includes page counts, topic counts, number of engineers per writer, time per page to edit, number of errors discovered per page, and so forth, ad nauseum. As soon as you start using a numerical measure, someone will try to optimize it, resulting in the kinds of dysfunction discussed above.

Who should measure

As much as possible, have writers make and use their own measurements. For example, page count and pages written per day can be useful in estimating the effort required to complete a job. In a delegatory management style, writers would keep track of these measurements and use them to come up with personal metrics they can use to estimate projects.

But, if *managers* measure those things, even if they do not intend to use them for comparison, they risk dysfunction. Therefore, I do not track this information; I let writers do that if it helps them, and I only get involved to help a new writer or when asked.

Once or twice, I have been asked to collect and report some numerical metric like pages per day. If you find yourself in this situation, my best recommendation is to find some analog in that manager's background (for former programmers it might be lines of code, for former marketers it might be number of ads or press releases) and use that to convince them that counting output is a bad idea. Nearly every discipline has some verboten measurement that is analogous to page counts; find it and use it. So far, this technique has helped me dodge that bullet.

How to use measurements

Austin identifies two uses for metrics, motivational and informational. He only advocates using motivational metrics when you can exercise full supervision. The catch is that nearly any metric can be used for either motivational or information purposes.

For that reason, I prefer to measure as little as possible and delegate measurement as far down as possible. In addition, I make it clear I will not use metrics as part of employee evaluation. Of course, I will have a discussion about milestones that are met or not met, that is inevitable. But, if writers have generated those milestones themselves, the discussion can be directed towards making better milestones, rather than towards reward or punishment based on making or not making a milestone.

There is a fine line here between management behavior that distorts and management behavior that enhances. You will not be able to make a perfect call in all situations, but if you have build a trusting, delegatory environment, the odds are that you will avoid the worst dysfunction.

Summing Up

Here are some thoughts to sum up the use of metrics.

- ► **Pay attention to what you measure:** People will assume that what you measure is what you value. A trivial example: If you measure pages produced per day and do not measure the quality of those pages, your writers will presume you care more about volume than quality.

- ► **Never use metrics in a PE:** As soon as you cite a metric in a performance evaluation, that metric will be optimized, not just by that person, but by everyone on your team. The metric will be optimized even if the optimization lowers the value of your product to customers.

- ► **Never use metrics to compare people or teams:** The same thing will happen in this case that happens if you use a metric in a performance evaluation.

- ► **Measure for information:** Certain measurements are necessary. For example, you need to keep track of milestones. When maintaining content, you need to keep track of error reports. And, you need to use these measurements for legitimate management purposes like estimating effort, identifying problems, and reporting status to the project team.

There is always the risk that these measures will be interpreted as evaluation measures, but you can limit this risk by minimizing individually identifiable measures, especially in reports to management, and by using the data strictly for its informational value.

► **Your credibility matters:** The fine line between measuring for information and measuring for evaluation is drawn based on the level of trust between you and your team. If you ever use an informational measurement in an evaluation or if there is a low level of trust for other reasons, you can be sure that your team will perceive any measurement as evaluative.

► **Let people measure themselves:** The most sensitive measures for writers are productivity measurements like page counts. These measurements can be useful for estimating effort and for judging when a project is off track. Let your team members estimate their effort using metrics they collect themselves. Delegate to them the job of measuring and acting on the results.

► **Do not punish *or* reward based on metrics:** It does not matter whether you use a metric as a basis to reward or punish. If you reward someone for producing more pages you will see the same result as if you punish someone for producing fewer pages. Either way, people will perceive page count as something you value and act accordingly.

► **Resist the pressure to measure productivity:** Measures like page counts are insidious to productivity. Your team may use page counts as part of their estimation process, but avoid using page counts as a productivity measure. If you are pressed by management to measure productivity using volume metrics, press back.

The bottom line for me is to measure only what you absolutely must, never use measurement as part of a performance evaluation, resist the pressure to report productivity metrics to management, and let writers manage their own metrics.

13

Localizing Your Documentation

It is no coincidence that in no known language does the phrase "As pretty as an Airport" appear.
— Douglas Adams

Unless your company sells American flags or some other truly local product or service, the odds are that you will need to adapt your documentation for one or more international markets or *locales*. As used here, a locale comprises a set of language and location-specific information that characterizes a group of users.

For example, if you want to market a product or service to people living in the French speaking part of Canada, you need to adapt your product to use the French language in its interface and documentation, and Canadian conventions when displaying information like the date and time, numbers, or currency.

At one time, adapting products for different locales was a mess. Countries defined unique national character sets for their language(s), and while there was some coordination through standards groups, the result was a hodgepodge of conflicting character sets.

Companies often handed localization of their products to subsidiaries in the target locales. This meant each locale had a unique product with significant technical differences, leading to a maintenance nightmare.

This mess has abated significantly, mostly thanks to *Unicode*, which is an international standard for character representation that has brought the vast majority of the world's writing systems under a single umbrella. Unicode is so important that if you learn nothing else from this section, at least remember to specify Unicode compliance in any tool you use.

This section will focus on what you need to know to localize your documentation. It is not a comprehensive explanation of localization and internationalization, but it will get you started.

Internationalization and Localization

When you adapt a product for use in multiple locales, you need to consider two dimensions: internationalization and localization. As a documentation manager you will mostly be concerned with localization, but you should understand both.

Internationalization

Internationalization (often shortened to I18N) prepares a product for use in more than one locale. A properly internationalized product contains support for the character sets, currency notation, date and time notation, and numeric representations needed for some set of languages. It provides a framework for users to plug in the translations and other information that will make the product work for a particular locale.

For example, if you manufacture speedometers for cars, an internationalized design would have a separate card for the speed markings. You could then print cards for miles per hour, kilometers per hour, feet per second, or whatever your customers needed, and add them either on the assembly line or after manufacture. The easier you make it to swap out the card, the cheaper it will be to support a new market that requires different markings.

The design that lets you easily swap out the markings is internationalization, while the creation of specific cards for different measurements is localization.

Characters and character sets

For documentation, the biggest question is usually what character sets are required for the languages you need to support. Each language has a set of characters used to represent that language. For languages like English, this is a relatively small number of characters (the 26 character alphabet (upper and lower case), punctuation, and few other symbols). But, for languages like Chinese, Japanese, or Korean, you need thousands of characters.

Computers do not directly process characters; instead, they process numbers. For the computer to handle characters, each character needs to be assigned a number that will represent it inside the computer. This mapping is called a *code set*. Until recently, the most common code set was ASCII, which contains the characters needed for most English words (the exceptions include words like résumé, which use characters with accents or other modifications).

Since ASCII only covers a small fraction of the world's languages, anyone trying to represent characters not in ASCII needed alternatives. This led to a proliferation of code sets, often overlapping and confusing. While schemes were developed to help manage multiple code sets, the situation was at best murky for many years.

Thanks to the Unicode Consortium [http://unicode.org] and ISO (the International Organization for Standardization) [http://iso.org], this babel of character sets has for most practical purposes been replaced by the *Unicode* standard. Unicode is a code set that encodes the characters for every language you are likely to need.

There is probably no good reason for most writing teams to use anything other than Unicode, usually in an encoding known as *UTF-8*. UTF-8 is backwards compatible with ASCII and is the default encoding for XML documents.

Local conventions

Beyond code sets, internationalization must support local differences in the representation of currency, dates, times, time zones, and formatting of numbers. Some of these things need to be handled during translation, but many can be handled with software.

For example, it is possible to represent a date or time in a computer using an international standard that applies worldwide. You can then translate that date or time into a string that makes sense in different places. This means that a program could store a date, say July 4, 1776, in a standard form and display it as "4 juillet 1776" in French or as "4 de julio de 1776" in Spanish.

Local conventions go beyond language. For example, even though Canada and France both use French, local conventions differ. The same thing occurs with Taiwan and China, which use the same language, but have significant local differences. The term *locale* identifies the combination of a language and a region, so you can have separate locales for Taiwan and China. You can also have separate locales for a country like Canada that uses two different languages.

Markup languages like HTML and XML use an attribute (`xml:lang`) to specify the locale for a particular document or part of a document. The attribute value is in the form `ISOLanguage[-ISOCountry]`, where `ISOLanguage` is an ISO 2 or 3 letter code (ISO 639-1 or ISO 639-2T) and `ISOCountry` is an optional ISO country code (ISO 3166).

For example, `fr` is the designation for the French language, without any particular country identified. `fr-ca` is the designation for the French language as used in Canada. The two part identifier allows you to account for the use of more than one language in the same country/territory (`fr-ca` and `en-ca` for French and English respectively in Canada) or the same language in more than one country/territory (`zh-tw` and `zh-cn` for the Chinese language in Taiwan and China respectively).

Generated text

Some standards, for example the DocBook standard, provide an additional feature that supports more generalized support for "boilerplate." In DocBook XML, this feature is called "gentext." Gentext defines localized phrases for standard phrases like: "Chapter," "Example," and so forth. When a publishing system, like the DocBook stylesheets, generates a document, it can plug in the gentext value for these standard phrases.

Gentext goes beyond simple translation to support parameter substitution. For example, the English gentext string for the title of an appendix is "`Appendix %n. %t`". This string tells the output transform to insert the appendix number in place of `%n` and the title in place of `%t`. If the appendix number is 6 and the title is "Some Extra Stuff," the transform will generate: "Appendix 6. Some Extra Stuff". If instead you set your language to Hungarian, the transform will pick up the following gentext string, "`%n. függelék - %t`", and generate output based on it.

Localization

Localization comprises those things you need to do to support one particular locale. A properly internationalized system will have well-defined processes for supplying locale-specific information. Consider the gentext capability. The DocBook stylesheets use the `xml:lang` attribute to select a file for the desired locale. That file is written using a simple schema that identifies the strings and their values. Gentext support is part of DocBook's internationalization capabilities. The individual files for each language, are localizations. Users get a set of localizations with DocBook, but they can also add localizations for new locales without touching the underlying internationalization support.

In addition to translated text, localization supplies locale specific information to handle display of numbers, currency, times, and dates. Localization also packages this information to be plugged into the product.

If the base product has been badly internationalized or not internationalized at all, you may need to do significant engineering to localize the product. Consider again our speedometer example. If the scale is engraved onto the surface of a sealed unit, you may need to replace the entire unit, rather than just slip in a new card.

When you are selecting tools, you should examine what it will take to localize for a new locale. This takes more than simply looking for that Unicode check-off item, though that is the first step. Just because a product will support Unicode does not mean that it can render bi-directional text, like Hebrew or Arabic, or that it can render languages like Chinese, Japanese, and Korean, which require large fonts and complex sorting algorithms. Ask for a demo of the product with content from the languages you need to support, and bring someone familiar with those languages to try them out.

Translation

Translation is probably the biggest and most expensive part of localization. Unless you are part of a large multi-national company that has in-house translation support, you will most likely need to deal with outside contractors, in different parts of the world, and you will need to manage their schedules and yours closely to meet your ultimate goals.

In my experience, translation is best done in the target country by native speakers of the language. There are some exceptions, but if you have read more than a few manuals for consumer products made in Asia, you know how bad a lousy translation can make you and your product look. This is not a place to save money by hiring a Spanish major from your local junior college.

What You Need to Know

Just as with every other aspect of your job, the localization situation in your company is going to be unique to that part of that company. To get your feet wet, you should start out by asking a few questions:

- ► **What are your responsibilities?** This may seem obvious, but you need to know what your company expects you to do. In some companies, local offices completely handle localization. In other cases you may be responsible for web sites, help systems, and printed documents in several local languages and be expected manage the entire process.

- ► **What locales do you need to support?** You also need to understand what each locale requires For example, if the locales include Taiwan and China, you might think that a single localization would work, since both places use Mandarin Chinese. However, the Chinese in China have simplified many characters and use different terms for many common things. A localization that is appropriate for China will not work in Taiwan, and vice versa.

- ► **How are localized products scheduled?** Sometimes you will need to support every language simultaneously at initial product delivery. Or, you may be able to stagger deliveries over a period of time. As you might guess, the latter will be much easier to manage.

► **How much content is being localized?** Do not assume that everything you write will be localized for all locales. Requirements vary; some deliverables may never be translated and others may only be translated into some languages. The sales and marketing people for each locale will usually decide.

Scheduling Localization

If you need to ship translated documentation at the same time that you ship English documentation, you have an intricate scheduling job in front of you.

In a typical product cycle, there is barely enough time to complete documentation without translation, and often you will need to update documents after features freeze. You never have enough time to simply shoehorn translation into the schedule as a single milestone between completing a document and delivering it. If you try to do that, you will have to either freeze English before the product development completes or re-translate pieces of your documentation as changes occur. Since you will not have the time to do the latter, you will end up with translated documentation that is out of sync with the English and out of date.

Table 13.1. Localization Production Cycle

Date	Milestones
01 October	► Feature A development freeze
08 October	► Feature B development freeze ► Feature A documentation content freeze ► Feature A content to translator
15 October	► Feature C development freeze ► Feature B content freeze ► Feature B content to translator
22 October	► Feature C content freeze ► Feature C content to translator ► Feature A content back from translator

The best strategy I know for managing this is to divide the translation job into smaller pieces and interleave it with the rest of the product cycle. Table 13.1 is an example of how you might interleave product feature freezes, documentation freezes, and translation deliveries.

Though simplified, this example illustrates the main points. Schedule your documentation development in small pieces that can be independently handed to translators as they are completed. If you are updating an existing document, first deliver any content you expect to remain unchanged, then schedule the rest based on when the features they cover freeze.

To make this happen, you need to have a close relationship with your translators and with whoever is responsible for back-end production of translated documentation. You cannot simply heave small pieces of your docs over the fence and expect to get back something coherent. Instead, your translators need to understand the overall schedule and work with you closely.

In addition, you need room in the schedule to send updates to your translators for the inevitable last minute changes, and you need to have an agreement with them on how they will be handled. You do not want to get in a situation where you pay for significant amounts of re-work just to change a few words in one section. At the same time, you need to make sure you do not saddle your translators with massive updates at the last minute.

This is an area where XML and modular documentation development techniques shine. If you develop content in small, self-contained modules, you will have much more flexibility in translation.

Another important tool is translation memory, which is software that stores translated content keyed to the original English text. When a previously translated English sentence occurs in a new or updated document, the software brings up the stored translation so it does not need to be translated again. A translation memory system is so important that I would not even consider using a translator who did not use one.

Minimizing Translation Costs

Translation will almost surely dominate the cost of localizing your content. Here are some suggestions for minimizing that cost:

- ► **Write consistent prose:** I am not a big fan of style guides; if you author one in house, it will be a huge time sink, and if you adopt one from outside, you will need to adapt it to your needs, which if you are not careful will be another huge time sink. However, consistent prose will make your content easier and cheaper to translate. You may want to look at the Plain English Campaign [http://www.plainenglish.co.uk], which has a set of guidelines for writing English that is easier to translate.

- ► **Limit vocabulary:** You do not need to turn your content into a Dick and Jane primer, but you also do not need to use unnecessarily complex language. Your objective is to convey information, not dazzle your audience with your erudition. The Plain English Campaign has some good thoughts on this topic, too. Just make sure you do not sacrifice clarity to make text more translatable.

- ► **Avoid unnecessary changes:** Translation memory systems store everything that has already been translated, so when you send your translators an update, they only need to translate the updated text. The less you change between updates, the less you pay.

- ► **Reuse content:** For the same reason, if you can reuse content, do so.

- ► **Write concisely:** You do not need to translate words you did not write. It can cost several hundred dollars per 1,000 words to translate content into just one language; every omitted word saves you money.

- ► **Use translation memory:** Make sure your translators use a good translation memory tool. Trados and Wordfast are probably the two best known commercial tools, but there are plenty of choices out there, including some open source programs.

- ► **Keep track of changes:** Know how much content has changed since the last release. It is not hard, or expensive, to do this. Standard translation memory tools can do the job, or you might try a tool like the Unix/Linux **diff** command. The idea is not so much to get an exact count as it is to make sure you are comparing apples when you look at your translation costs.

► **Do not skimp:** Bad translations, even if "cheap," are not economical. The first time you work with a translation house, start them with a small job and get an independent evaluation of their work from a native speaker of the language in the target country.

► **Only translate what you must:** Some documentation in some markets is not worth the expense of translation; maybe the market is too small to justify the cost or the target audience is able to work in English or whatever your original language is. For example, airline pilots trained in the US by Boeing should understand enough English to work with checklists and other technical documents in English.

14

Single Sourcing

A place for everything and everything in its place.
— Isabella Mary Beeton *The Book of Household Management, 1861*

A writer's workload is directly correlated with the volume of content that writer must manage. Stripped down to its essence, single sourcing is a strategy that minimizes the volume of content by reducing duplication.

An obvious example is boilerplate. If you have a section about typographical conventions in every printed manual, there is no point rewriting that section every time you begin a new book. Instead, you reuse that section everywhere you need it. This is the simplest and most common form of single sourcing.

Most good writing groups will share and reuse boilerplate. If they are really good, they maintain that content in a central location that meets the minimum requirements for a Content Management System (see the section titled "Storing and retrieving content" (p. 217) for a list). The best maintain all of their content in this manner.

Groups that take full advantage of single sourcing go much further, carefully planning their entire documentation set with single sourcing in mind, and exploiting every opportunity to eliminate duplication.

The potential is large. If you analyze the content accumulated over the years by a long-standing documentation group, the amount of unnoticed duplication can be astounding. In one group I worked with, we analyzed their existing documentation set for duplication and discovered hundreds of sections that were duplicated, either verbatim or nearly so. Some of these sections were repeated in as many as seven different locations.

Over the years writers had been reusing each others text, which was good, but they were not maintaining the reused text in a common place, which meant that as soon as they made a copy, they had to maintain multiple instances of the same text. And, in many cases at least one of the copies was out of date or inaccurate.

A good single sourcing strategy will reduce the volume of content that needs to be maintained, store common copies in a single place, and provide easy access for users. The result minimizes the amount of content and makes what remains more consistent.

There is another dimension of single sourcing, delivery of content to different media. For example, you might take a set of procedures and deliver them as a printed manual and also in a help system. I will refer to this type of single sourcing as *content repurposing* and the first type of single sourcing as *content reuse*. Let's look at strategies for each dimension in turn.

Content Reuse

A good content reuse strategy begins with the organization of your content, and paradoxically, the best organization will minimize, not maximize, content reuse. I like to think of it in terms of the old saw, "A place for everything and everything in its place." A good organization will have one logical place for any given piece of content and will rarely require you to reuse content in more than one place.

The reasons for minimizing reuse are simple. First, every time you reuse some content, you give your users yet another place to look when they search on line or in a printed index. If you have the same content in several different places, your users end up jumping around among those places, trying to figure out which one they should use. Having one, authoritative place for any particular module will simplify their search and avoid confusion.

Second, even with highly structured methodologies, content reuse is not free. For example, if you aggressively drive out duplication, you will inevitably find places where the same content *almost* fits in two or more locations, and you need to decide how to handle the situation. When this happens, you have three choices.

1. Maintain separate copies, one for each location

2. Maintain one copy and edit that copy to fit both locations

3. Eliminate one, or both, of the locations

All three of these choices will cost something. If you take choice one, which gives up on any kind of reuse, you will end up with more content to maintain. That does not make this a bad choice as long as the cost of maintaining that content is less than the cost of using one of the other options.

Choice two is classic reuse; you will have some additional work making the module work in both locations, and you will have some additional work over time maintaining that independence.

Choice three eliminates duplication and reuse. If you can eliminate one of the situations that used the content, you will both eliminate the duplication and reduce your overall content. When it works, this is the most efficient of the three and ought to be your first choice.

However, too few writing groups follow this approach. Most try to maximize reuse, heading straight to choice two. They may not even consider choice three. In fact, given that it is easy to create a metric to measure reuse, but difficult to create one to measure where you have avoided the need to reuse, the metrics themselves will create a bias towards choice two.

To make things worse, the typical Content Management System makes reuse easy. Just mix and match modules, push a button, and poof, you have a new deliverable.

If unchecked, these biases will leave you with a lot of unnecessary reuse. You can argue this is not a big deal, but even when well structured, a heavily reused module will take more maintenance than one that is used in just one place. In addition, it will needlessly increase the bulk of your deliverables. Both of these factors decrease efficiency.

I recommend a *minimalist* approach to the content reuse side of single sourcing. First, create an architecture that minimizes the need to reuse content. Second, use a writing methodology that makes reuse easy when you need it. And, you will need it, because it is not possible, nor desirable, to completely eliminate reuse.

You will find places where reuse is appropriate; for example, legal notices, typographical conventions, glossary entries, and other boilerplate. In addition, you will find places where reusing content is preferable from a usability perspective. After all, you probably do not want to point a user to a far corner of the documentation set for a short task or reference item. In those cases, it usually makes sense to place a copy in line.

That said, if you take the time to architect your content and deliverables carefully, you will probably end up with much less content reuse than you expected.

Content Repurposing

Repurposing directs content to multiple delivery vehicles. A standard example would be to take the content of a help system and make it available in printed form for people who prefer a book. Another example would be to take a printed book and make the information available on a web site.

I worked with one organization that maintained the same words in two different forms so they could deliver print and help. When there was a change, both copies had to be edited. Few organizations can afford to maintain two copies of anything but the smallest amount of content, and in fact, this particular organization has since moved to a single sourcing model that avoids the duplication.

Most organizations begin repurposing when they decide to take some existing content, typically print, and deliver it in some other medium, typically web or help. I have worked with organizations that actually maintained two copies of the same content, one for print and one for their help system. For these organizations the challenge is usually not technical; it is straightforward to take almost any content and repurpose it. Instead, the challenge is how to structure and write content so that it will work in both the old and new media.

An existing book repurposed for the Internet without any editing will almost always be harder to read than the print equivalent. As discussed in the section titled "Developing Content for the Internet" (p. 200), people read content differently on the web than they do in print. Content written for the web may seem choppy and disconnected to someone reading it in print, and content written for print may seem windy and poorly organized to someone reading it on line.

This poses a real challenge. There really is no good way to make most content work perfectly in both print and on the web. You will need to compromise somewhere. Most organizations compromise in one of two ways.

- ▶ **The Book-ware model:** This model can make good sense when the original content was in book form and where there are limited resources for editing the content to make it more suitable for repurposing.

  The upside of this model is that it requires less work when going from books to online. The downside is that web readers will find it harder to use your content.

- ▶ **The Block-ware model:** This model makes sense when you are starting from scratch or starting from content, such as a help system, that is already in modular form. It may also make sense if you are starting from content in book form, and you have the resources to adapt that content for a modular methodology.

  Upsides are that you will have an easier time single sourcing the content, your users will find it easier to use your content, and if done right the printed form will be usable, if not ideal. Downsides are the amount of work needed if you start from a book model and the danger of going so far into modularization that the content becomes choppy and hard to use in any medium.

If you are starting from scratch, I recommend looking carefully at the Block-ware model, even if your primary deliverables are books. When well executed, this model will give you good printed material, excellent online content, and support for both reuse and repurposing.

Managing Technology

15

Living with Technology

If the only tool you have is a hammer,
you tend to see every problem as a nail.
— Abraham Maslow

When I first became a documentation manager, I thought I was leaving the engineering world. Even though the term is "technical" documentation, I thought of it as a non-technical discipline. Not true then, and not true now. Technology has become a central focus for documentation managers. If you are not comfortable dealing with technology, you will find managing technical communication difficult if not impossible.

This does not mean you need to be an engineer or programmer, but it does mean you need to understand what technology can and cannot do for you, and to the extent you are not fully comfortable with technology, you need to find competent people you can call on for help.

In this part of the book, I start with some observations about living with technology. You can think of them as rules of thumb that you should keep in mind as you tackle technology issues. Then I look at acquiring technology, with an emphasis on the ground work you need to do before approaching vendors, followed by a chapter on building a business case.

The next three chapters look at the three most important technologies in this area: XML, the Internet, and Content Management Systems. I look at these in depth because they are the latest trends as of this writing, and because any one can improve, or destroy, your team's productivity. I round out this part of the book with a discussion of common pitfalls and how to avoid them.

Rules of Thumb

- **Not every problem can be solved with technology:** If you spend enough time with technology vendors, or if you work for a technology vendor, it is easy to assume that technology can solve any problem. Instead, expand your toolkit beyond technology to consider the structure of your content, your processes, and the skills of your team whenever you try to solve a problem or make an improvement.

- **Technology must follow, not lead:** Your job is to produce a product, not use technology. The latest, hottest technology is worthless unless it helps you do your job.

  Consider content reuse. If you read a few vendor white papers and go to some conferences, it is easy to assume that you need to go out right now and buy an XML Content Management System (CMS) to maximize reuse. Before you succumb to that urge, consider the following: Do you know what the number one productivity problem is for your team? If it is not reuse, or if you do not know what the number one productivity problem is, then you are letting the technology lead.

- **Keep it simple:** To paraphrase Einstein, "Technology should be as simple as possible, but no simpler." Technology is nearly always developed by people who want to satisfy as many user needs as possible. Unless they exercise tight discipline, developers are almost guaranteed to produce products that contain every feature they imagine anyone might ever use, plus a few they thought were cool. The result is that most products contain a laundry list of features, most of which you will not need.

  The best way to combat this tendency is to be clear about what you need, evaluate products based on those needs, and find a way to turn off (or at least hide) the features you do not need. The best technologies are like the best managers, they provide an environment where you can focus on producing a quality product with minimum distractions and maximum support. Extra features are unnecessary distractions; get rid of them.

▶ **Take advantage of the broader community:** Any useful technology will have a community of users outside, and possibly inside, your company. You should look for news groups, wikis, message boards, user groups, and mailing lists for any tool you are using or considering using. If you cannot find a group of people actively using a technology, stay away.

Once you are linked into the community of people using the same technology, you will have access to information and people who can save you a lot of time and frustration. A great example of this is the community of people who use the DocBook XML standard. The DocBook mailing lists [http://docbook.org/help] are regularly monitored by experienced users, including the people who develop and maintain the DocBook standard and its stylesheets.

▶ **Avoid orphans and old-timers:** Avoid products that no one is using or are past their sell-by date. You want a product that is mature and evolving, with a strong and growing user base. This is not hard to spot; just spend a little time on the web. A quick review of the last few months of activity on any user group will tell you a lot about the viability of the technology. Look for an enthusiastic group of users that responds quickly, more discussion about how to take advantage of the product than discussion about avoiding problems, active participation by the technology's developers, and productive discussion about future releases.

▶ **Avoid newborns:** While it is pretty obvious why you should avoid the obsolete and orphans, it is a lot harder to resist new technology. But, even if it looks like it will cure cancer, feed the poor, and eliminate bad breath, you do not want to be an early adopter unless you are experienced in adopting new technology, have the internal expertise and time to shake out problems, and have a real need for that specific technology. Otherwise, let someone else take the hits and wait for the technology to mature.

► **Garbage in, garbage out:** This one applies particularly to Content Management Systems, though it also applies to other technologies. If you do not have well-structured content, no technology will save you.

Just as important, if your processes are bad, automating them will not make them any better. In fact, it will probably make them worse. If you are a lousy driver and buy a sports car, the car will not make you a better driver; in fact, it will exacerbate your bad habits. In the same way, technology can accelerate your process and make it easier to create content, but it will not improve the quality of either.

16

Acquiring Technology

As few as 10 percent to 15 percent of
Enterprise Resource Planning (ERP) implementations
have a smooth introduction that delivers the anticipated benefits.
— A. Blanton Godfrey

At some point you will need to acquire technology to support your work. The Godfrey quote is about as pessimistic as I have seen, but even more optimistic reports[1] suggest failure rates as high as 37% for software projects. Projects that depend on deployment of software are complex and subject to what often seems like an infinite number of ways to fail. This chapter looks at how you can improve the odds when you acquire and deploy technology.

Defining the Problem

Before you begin the process of acquiring new technology, you have some legwork to do. Take a step back and consider your underlying motivation for change. Usually the motivation is to improve the product

---

[1] Scott Ambler[1] suggests failure rates for "traditional" and Agile software development projects as 37% and 28% respectively.

or to improve productivity. These are worthy goals, but it is easy to be seduced by a slick sales presentation or a whizz-bang white paper. You need to dig deeper.

What specifically are you trying to accomplish? If you want to improve the quality of your product, are you trying to improve accuracy, completeness, writing quality, searchability, or visual appearance. If you want to improve productivity, are you trying to shorten production cycles, remove overhead, speed localization, reduce headcount, or something else? Be specific. If you are trying to shorten production cycles, specify how much shorter you need them to be and what the benefit will be if you shorten them by that amount. If you are trying to improve accuracy, then by how much and how are you going to measure progress.

Be specific about *what* you want the end result to be, but avoid defining *how* you are going to get there. This is tricky; it is very tempting to state something like: "all of our documentation content will be authored in XML." That defines an end result, but it does not define a tangible improvement (some would say it defines a step backwards). Even though it reads like a *what* statement, it is really a *how* statement, and even worse, it does not offer a reason for doing this. A better statement would be something like the following:

> Through an analysis of our documentation, we have discovered that about 20% of our content is redundant. That is, we have exact or near exact duplication of content in different deliverables. While authors are aware of some of this duplication, it is difficult for them to coordinate reuse of content so that they can eliminate this redundancy.

> Our objective is to eliminate 80% of that redundancy within 18 months and 95% within 24 months without adding more than 5% overhead for writers. We will measure the elimination of redundancy by repeating the initial analysis, and we will measure writer overhead by interviewing writers.

This "Problem Statement." identifies and quantifies the problem, establishes a goal, and describes how and when progress will be measured.

It says nothing about how the objective will be reached. In fact, it does not even demand a technology solution. Maybe simple awareness of the problem and better communication are all you need.

Defining Requirements for a Solution

The requirements specification is the next important step in the process. It takes you further down the road towards a solution by defining a more detailed set of objectives. This is also the point at which you begin to narrow in on the kind of solution you need, even though you still are not choosing technology.

This step is critical. In my experience, most projects fail because the requirements specification was either skipped or botched. Here are a few guidelines to keep in mind:

- ► **Document your requirements:** You will need to present your requirements to your management, vendors, writers, and other affected parties. Formally documenting requirements will force you to give this part of the process serious consideration.

- ► **Start with what you need, not what you want:** I have participated in more than a few efforts to gather requirements where managers would tell me they *needed* features like workflow management or detailed tracking. Yet, when I looked at their current processes, I could not even find placeholders for those activities. That does not mean that a feature like workflow management is useless, but if you are not already at least trying to track workflow, then workflow management is not a "must have" feature.

- ► **Document the reason for every requirement:** There should be an anticipated benefit for every requirement you list. If you can attach a dollar benefit, great, but even if you cannot attach dollars, you need to have a good reason for every requirement.

- ► **Prioritize your requirements:** You can't always get what you *want*, but if you know what you *need*, and can prioritize those needs, then you can make the tough choices.

  I like to prioritize requirements as an ordered list. Then I can compare any two requirements and know which is more important. This means avoiding priority buckets like: "Must Have," "High

Want," "Nice to Have," and so forth. If you use buckets, you will end up with too many "Must Have" features and you will have a hard time when you need to cut features from the list – something you always have to do at some point. I sometimes use buckets for a management presentation, but only after creating a prioritized list.

- ► **Do some outside research:** No matter what you are trying to do, the odds are that someone has tried to do it before. Get on the web, look at what others are doing, and make some friends. It will not take long to get a sense of which requirements are feasible and which are pipe dreams. You will also get a head start on solutions. However, do not spend too much time looking for solutions; you do not want to limit your options this early in the process.

- ► **Do your work in the open:** The more closed the requirements gathering phase is, the harder it will be to get people to buy in to the result. And, the fewer people you have looking into the requirements, the more likely it is that you will either leave out something important or give something unimportant too high a priority. I have seen more than a few requirements gathering efforts fail because people felt left out or because the effort was so narrow that important information was overlooked.

Gathering input

Defining requirements should be interactive. It is very tempting to hide out in a cave with a few like-minded people and pound out an idyllic version of the world as *you* would like it to be. If you have tried to run a wide open process for gathering requirements, you can understand why that is an attractive idea. As soon as you open up a requirements gathering activity, you will be inundated with proposed requirements, each one absolutely essential to the very existence of the person who proposed it.

Your job is to embrace this onslaught, not avoid it. The idea of having an open process for gathering requirements is not to give everyone exactly what they think they want. It is to give everyone a voice in the process and to give everyone a chance to hear everyone else's voice. You then need to take that onslaught of ideas and turn it into a specification that reflects the real needs of your user base and can be implemented on budget and on schedule.

The best way to manage all this input is to focus your attention and the attention of everyone giving you input, on what you need, why you need it, and what the priority is for each thing you need. Do not just ask for wish-lists, get them directly involved in prioritizing needs. If you just let them give you ideas, but do not get them involved in narrowing down choices, they will never be happy with the result. But, if you get them involved in each step of the process, they are more likely to buy into the result.

Defining use cases

Let's take a look at the process in more detail. What we are doing here is decomposing a set of very high level objectives, like "Reduce duplication of content by 32%," into a set of requirements for technology, typically software, that will give you the tools you need to achieve those objectives. But, you are not just defining a set of features; you are building a complete picture of how you are going to meet your objectives.

Use Cases provide one way to build this picture. According to Wikipedia, "Use cases describe the interaction between a primary actor – the initiator of the interaction – and the system itself, represented as a sequence of simple steps." When well defined, they can act as the requirements for your system; that is, if the technology you adopt enables actors to complete the interactions defined in your use cases, then that system should meet your requirements.

To illustrate the importance of use cases, I will use the example of an organization, which will remain nameless, that acquired a well-known Content Management System because management had the idea that the kind of modular documentation this CMS supports would improve productivity and quality. Unfortunately, they skipped right past use cases and requirements specifications, and tried to deploy the system into the middle of the organization's existing activities.

The fatal flaw, or to be more accurate one of several fatal flaws, was to completely ignore use cases. This led to two serious problems:

- ► The interface between the CMS and the back-end systems (print and web production) was never considered. As a result, hand-offs required a series of complex manual steps.

- ► Use cases for content reuse were never explored. As a result, even though the CMS had features that supported content reuse, it was set up in such a way that those features were hard to access and nearly unusable.

Therefore, a system that might have improved productivity and quality instead harmed both. And the harm was largely independent of the particular technology they chose. It is likely that the CMS they chose could have handled their needs, had they defined their use cases completely and in the proper context. The lessons here are to look at your use cases, processes, and content architecture as part of the requirements definition process.

Writing the Requirements Specification

A Google search for "Requirements Specification Template" yields over 10,000 hits. After looking through a fair number of them (more than 10, less than 10,000), I do not think you need yet another template from me. Instead, I will give you an annotated outline that covers the most important elements of a Requirements Specification. You can turn this into a template, or you can look at some of the ones available on line and pick one that matches your exact needs.

Introduction

This section introduces the document and serves as an Executive Summary. In fact there is nothing wrong with calling it that.

Once someone has read this section, he or she should have a high level understanding of what the objectives are, why the project is necessary, who will be affected, what the scope of the project is, and what the benefits are. Major sub-sections include:

- ► **Purpose/High Level Objectives/Scope:** What are you trying to do? Why are you bothering? What is the scope of the project (what activities and people will be affected).

- **Conventions/Terminology:** The usual compendium of typographical conventions and specialized terminology. If you have a lot of terminology, you may want a glossary at the end instead.

- **Audience:** Who is the audience for the specification? This will normally include the people who will build, deploy, and maintain the system, the people who will use the system, and the people who will fund it.

- **References:** Standard reference section, though again, if it gets big, stow it at the back as a bibliography.

Overall description

This section takes things a step deeper, but keeps the discussion at a broad system-wide level. Once someone reads this section, he or she should understand the project objectives, how the proposed technology will fit into your current environment, what the assumptions, constraints and dependencies are, and what the approximate budget is. Major subsections include:

- **Objectives:** This section defines the overall objectives for the project. It should include an expanded description of the problem you are trying to solve, as well as the more detailed objectives that would define a good solution.

- **Environment:** This defines anything relevant about the environment the new technology will exist within. It might include a definition of the technologies that have an interface with the new technology, the software and hardware environment, and possibly also the organizational environment, if that is relevant.

- **Assumptions/Constraints/Dependencies:** You may need separate sections for each of these three. They are pretty much self-explanatory, with the one caveat that it is generally better to state an obvious assumption, constraint, or dependency than to assume it will be understood. The only safe assumption you can make is that any assumption, constraint, or dependency that you leave out because it is too obvious will come back to bite you.

- **Budget:** What are you willing/able to pay, both in dollars and in staff time. In some cases, you may not be able to complete this until you have a better idea of your needs, and you may want to omit this information when you share it with some parties, for example vendors, but, it is important to know and document how much you can spend.

Use cases

Not all requirements specifications identify use cases, some methodologies do not use them at all and others document them separately. I believe they are important, and that they should be documented, but I consider it unimportant where they get documented. I discuss them here because they are part of the overall process of acquiring technology.

There are at least as many templates for use cases as there are for requirements specifications, but the most widely quoted templates seem to be those derived from the work of Alistair Cockburn.[11] You probably will not go wrong if you use his template or one derived from it, and if you get deeply into use cases, you should check out his book on the topic. [10]

Here are the major parts of a use case. For this part of the discussion, let's consider the example of a highly automated restaurant, where we replace some of the waiter's functions with a computer system. The use cases here will define the functions of the computer system that replace ordering and bill paying.

- ► **Use case name:** A good use case name is one that is short, starts with an active verb, and gives the reader at least a notion of what the use case is about. For our example, you might have use cases with the following names: "Order Food," "Send Back Order," and "Pay Bill."

- ► **Actors:** The actors are the people and systems that interact in this particular use case. Many definitions add the idea of a "Primary Actor," who is the person who is trying to do something. For our three use cases, the primary actor is the customer. Other actors might include waiter, cook, busboy, hostess, and bartender

- ► **Pre-conditions:** A pre-condition defines the state of a system before the use case begins. In our example, a pre-condition for "Pay Bill" might be, "Customer has completed meal." Completion of some other use case might also be a pre-condition. For example, a "Cook

Food" use case probably requires completion of the "Order Food" use case.

► **Post-conditions:** A post-condition defines the state of a system after the use case completes. In our example, a post-condition for "Order Food" might be, "Cook has received all information needed to begin cooking."

► **Description:** The description is a high level explanation of the activities the use case comprises. For our example, the "Pay Bill" use case could be described at a high level as, computer displays bill for customer, customer swipes credit card, system prompts customer for a signature, customer signs a touch sensitive display, and system prints receipt for customer.

You could imagine a variation where the customer uses cash instead of a credit card. That variation might be described within one use case, or as a separate use case. Either is valid. If you choose separate use cases, you could have use cases named "Pay Bill with Credit Card" and "Pay Bill with Cash." If you keep both alternatives within one use case, you would use two "Flows" as described below.

► **Flows:** The other way to handle the "Pay Bill" case described above would be to have two different "Flows" within one use case. A flow is simply a different path through the same case; if you are familiar with computer programming, you can think of a flow as the two paths the code would take after an "if" statement.

Some templates use flows to document exception conditions, like a credit card being rejected. Others use a separate exception section. Either works fine; what is most important is to make sure you handle exceptions.

► **Assumptions:** A list of assumptions. For example, you might assume in the "Pay Bill" use case that the restaurant will not let customers split bills or that the restaurant does not use discount coupons.

► **Other considerations:** This can include non-functional requirements (performance, quality, etc.), issues, and other notes. For our restaurant example, we might want to specify that the time elapsed between the customer swiping the credit card and the system prompting for a signature cannot exceed 5 seconds without giving the customer some indication that the transaction is in progress.

Functional requirements

A list of requirements is an alternate way of defining what you want in a system. If your use cases are specific and well defined, then they may very well serve as requirements. But, if they are cast more loosely, you may want to have a set of functional requirements to tighten things up.

Since our objective here is to define what we want some technology to do for us, I think use cases are a better and more intuitive method than requirements. Requirements tend to drag you into over-specification. If you are actually implementing the system, you may want functional requirements instead of use cases, but for specifying what you want to a third party, I suggest you start with use cases.

Non-functional requirements

These are requirements that are peripheral to the specific functionality of the system. This does not mean that they are unimportant. Examples include: Documentation, Performance, Environment, Security, Localization, Software Quality, User Interface, Legal, and Budget. Even if you define use cases, you will likely have non-functional requirements that cover all of your cases.

Going back to our restaurant example, possible non-functional requirements might include: the system interface and on-screen documentation must be localized into English, Spanish and Chinese; the restaurant owner interface must be password protected; or the system documentation must include installation instructions, administrative instructions, and online help for customers.

Working with Vendors

Once you have documented requirements, you are now ready to talk with consultants and vendors. Here are some guidelines for this step.

▶ **Document what you need before you talk with a vendor:** Make sure you have a thorough set of written requirements before you talk with vendors. Share them with the vendor before you meet, and ask them to present to you how their offering satisfies those requirements. If you do not do this, you will get a standard sales pitch that may or may not address your needs. And even worse, you risk getting sidetracked into discussions that will without a doubt focus on what the vendor does well, rather than how they can solve your problem.

► **Talk to more than one vendor:** Meeting with vendors, while it is a nice diversion, takes time and effort. Your inclination will be to meet with fewer than you should. Resist that inclination.

► **Talk with people who use the technology you are considering:** If you are a member of the Society for Technical Communication (STC) or other interest groups, ask around. You should be able to find someone who has experience with the technology you are considering. If you cannot find anyone, consider that a red flag.

If you are working with anyone but a startup, you should able to find someone who is using the technology. And, unless you are really technically savvy and know exactly what you want, stay away from startups. You may miss some good technology, but unless you know what you are doing, you could also end up in an expensive mess.

► **Multiply whatever you pay for software by 2.5:** No matter what you spend on software and hardware for a technology, you will spend several times more to get it running in your environment. This includes training, maintenance, initial design, and customization. Somewhere between two and three times the cost of software is about right.

That does not mean your cost will be zero if you use open-source software. You may save initially with open source software, but you still will have additional costs. If you use open source software, take the cost of a commercial equivalent and multiply that by two for a rough estimate of your total cost.

► **Look at any claims for reuse very carefully:** Reuse can yield real benefits, but claims that system X will increase reuse by Y percent are usually worthless. Reuse depends on understanding your content very well and then planning for reuse. Technology can make it easier to reuse content, but it will not do the work for you. Also, every organization has a different potential for improvement in this area and no simple formula can calculate possible gains.

17

Building a Business Case

Drive thy business or it will drive thee.
— Benjamin Franklin

A business case documents the justification for pursuing a particular project. It outlines why the project is necessary, what it will cost, and what the benefits are from a business point of view. For documentation managers, nearly all business cases fall into one of three categories:

- **Justification for your existence:** This is basic; you should be prepared at any point to make the case for why your organization is worth the expense. You may not be called on to provide a detailed written business case for your ongoing existence, but you should be able to explain to management, using real numbers, why they should continue to fund your team.

- **Justification for hiring:** If you are trying to build an organization, you should have a business case for hiring, including an explanation of what the new person(s) will do and how their work will reduce costs, increase revenue, or decrease risk. Again, you may not need to document this in detail, but you should be prepared to justify your requests.

► **Justification for technology purchases:** You are most likely to first bump into detailed business cases when you try to justify a technology purchase, and this is where I will focus most of my attention. To justify a technology purchase, you will probably need to create a written business case and justify the expense based on cost savings or increased revenue.

Business Case Basics

Business cases will always contain at least the following information:

► **Business need.** A description of the problem you are trying to solve or the opportunity you are pursuing. A typical business need for a documentation manager might be to increase productivity for the writing team.

► **Proposed solution.** A description of the solution you are proposing. For example, to increase productivity, you might propose a single sourcing Content Management System that would improve productivity by reducing the amount of content that needs to be maintained and automating delivery of content to print publishing systems and the web.

► **Cost.** A description of what the solution will cost. In our example, the cost would include software, hardware, consulting, ongoing maintenance, and training.

► **Benefits.** A description, in financial terms, of how the solution will benefit the corporation. Typically, this means describing how the solution will increase revenue or decrease costs. In our example, the benefits could include reducing the number of pages of content that must be authored and maintained, and reducing the effort and time required to publish content.

The business case addresses the financial impact of a project, and as such must be written in those terms. While upper management may sympathize with the need to improve processes or documentation quality, they will only fund a project if you can show that those improvements will either increase revenue/profit or decrease expenses.

Profit Centers and Cost Centers

Upper management has two primary considerations when evaluating a business case: the financial cost and the financial benefit. In practice, the kind of business case you write will depend on whether your team is seen as a *Profit Center* or a *Cost Center*.

▶ **Profit center:** A profit center brings in more revenue than it spends. Product development organizations are typical profit centers. Even if you do not derive revenue directly from documentation, if your content is an integral part of a product or service and you report into a development organization, you will probably be perceived as part of a profit center.

▶ **Cost center:** A cost center spends more than it brings in. IT, HR, janitorial services, and so forth, are cost centers. While most cost centers provide necessary services, they bring in little or no revenue. In many organizations, documentation is seen as a cost center because it does not directly bring in revenue.

Documentation groups are odd birds; depending on the organization, they may be seen as either a profit or cost center. They rarely develop a product that is sold separately for a profit, and when they do, that product is typically a book that brings in negligible revenues. However, documentation is almost always an integral part of a product or service that does generate revenue.

Whether a documentation group is considered a profit center or a cost center depends in large part whether management perceives documentation as an integral part of the product that helps differentiate it from competitors or as a commodity that should be acquired at the lowest possible cost.

Building a cost center business case

If you are part of a cost center, management will expect you to write a business case that reduces expenses or avoids risks. There are several rationales you can use do this:

▶ **Direct cost reduction:** In this case, you show that a solution that costs X will directly reduce other expenses by Y, where of course,

Y is larger than X. For example, replacing a computer system with one that costs less.

- ► **Indirect cost reduction:** In this case, you show that a solution that costs X will result in cost savings in another organization by Y. For example, implementing a Frequently Asked Questions (FAQ) list in an online forum so that you can reduce support calls.

- ► **Improved productivity:** In this case, you show that a solution that costs X will improve your team's productivity by Y. For example, implementing a single sourcing system that reduces duplication of content.

- ► **Decreased risk:** In this case, you show that a solution that costs X will reduce some risk to the corporation that could potentially cost significantly more than X. For example, implementing a quality assurance (QA) program to reduce the possibility of a lawsuit for negligence.

These rationales range from easy to nearly impossible to quantify, in the order listed. This does not mean that you cannot write a successful business case that promises to reduce risk, but you will have a harder time coming up with hard financial numbers.

If you are trying to make a business case using a hard to quantify rationale, consider using more than one rationale. For example, you might argue that a single sourcing system will improve productivity by reducing overall content, reduce risk by improving the accuracy of your content, *and* reduce direct costs by upgrading your computer systems. The direct costs will be easier to quantify, and the other arguments will add weight to your case, even if you cannot come up with hard numbers.

One big downside of being perceived as a cost center is that management will always look for direct cost reductions. If your organization is expanding because of increased work, then you can make a successful business case that shows a slower growth rate. But, if your workload is steady or decreasing, management will expect your business case to decrease your overall budget unless you can make a strong case for one of the indirect benefits.

Building a profit center business case

If you are a profit center, or part of one, management will most likely be open to a wider range of business case rationalizations. You can still use the rationales listed above, but you will also have a new set based on increasing revenue/profit:

- ► **Increased revenue/profit:** In this case, you show that a solution that costs X will increase revenue by Y and profit by Z. For example, you might show that including a training DVD with your product would increase its value and justify a higher price.

- ► **Indirect profit:** In this case, you show that a solution that costs X will reduce a product/service related cost by Y, leading to increased profits. For example, you might propose delivering documentation on a CD or via the Internet to reduce shipping costs.

- ► **New product/service:** In this case, you show that the introduction of a new product or service at cost X will bring a return of Y. For example, you might propose developing training based on your content, and offering that training as a separate service. This is the most complex type of business case and will invite the tightest scrutiny. At the same time, it offers the greatest potential reward, both for the corporation and for your team.

While you get more flexibility as a profit center, you will need to do more homework. Every organization has criteria for evaluating a business case for *Return on Investment* (ROI). Your management team should be able to give you guidance on what they are looking for in terms of ROI and in terms of what they want to see in a business case.

How to look like a profit center

If your team is perceived as a profit center, you will have more options and leverage when you write a business case. While some functions are clearly cost centers, documentation can live in either world. By default, most management teams will see documentation as a cost center. However, this perception can be changed.

The biggest determinant for how management perceives documentation is where they live in the organization. If documentation lives in a cost center, then the odds are pretty good that management sees it as a commodity. If it lives in a profit center, the odds are better that management sees it as integral to that center. However, sometimes documentation will end up in a profit center, but be perceived by management as a cost center.

In practical terms, this means that wherever your organization lives, you need to shape management's perception so they see documentation as integral to the corporation's profits. Here are a few suggestions for doing this:

- ► **Participate actively in your projects:** In Chapter 10, *Project Planning* (p. 101) I recommend staying closely involved with the projects you supply documentation for. The closer you are tied to project schedules and deliverables, the more integral you are to the project. If you simply toss a book over the fence at the last minute, you will look like a detached commodity supplier. If you deliver content with every development cycle, you will look like an integral part of the product.

- ► **Talk the talk:** Talk about your work as part of the project, rather than a service or adjunct. Consider the difference between, "we write documentation for the XYZ product," and "we make it easier for customers to use our XYZ product." The latter outlines the benefit you provide and identifies you with the product, while the former positions your work as an add-on.

- ► **Use the "right" vocabulary:** Speak in the vocabulary of a profit center. This means using terms like ROI, rather than cost avoidance, even if you are avoiding cost. I realize purists may cringe, but profit centers use terms like ROI and cost centers use terms like cost avoidance.

- ► **Add value:** In addition to the basics, contribute to the project in other ways. Chapter 3, *Power and Influence* (p. 17) describes how a team became a more integral part of a project by going beyond the expected. The more you are involved in the project, the more you look like you are part of a profit center.

Writing the Business Case

Your management will dictate the exact form for a business case. If they do not have templates, ask for copies of successful business cases for other projects. However, no matter what the exact form is, any business case will contain the following elements:

Business need

The business need defines the opportunity being pursued or the problem being solved. It baits the hook. After reading this section, a manager should be thinking that you have identified either a compelling opportunity or a serious problem that deserves attention.

If you are pursuing an opportunity, you will most likely describe an unserved or badly served market. For example, if you are proposing a training product, you might show that there are no equivalent offerings from your company or competitors, and then document a need for such an offering.

If you are solving a problem, here is the place to describe the problem in its full glory and outline the consequences if the problem is not solved. For example, you might describe how the support team is receiving a large number of calls from customers who are confused about how to optimize some aspect of your product. You can then describe the consequences of this in terms of the cost of those calls and the level of frustration expressed by customers.

Proposed solution

The proposed solution describes how you propose solving the problem or pursuing the opportunity. If the previous section baits the hook, here is where you start to reel in your catch. After reading this section, a manager should be convinced that you have found the best possible approach, and be ready to sign on, once he or she understands the financials.

You should spend some time in this section explaining alternative solutions. Management will want to know that you have considered more than one way of approaching the problem or opportunity, and that you have good reasons for the solution you have chosen. This may mean you need to go into some financial depth on other solutions to show why they were rejected.

Cost

This section goes into depth on the cost of your proposed solution. For a technology proposal, you will need to cover at least the following topics:

- **Acquisition costs:** This includes everything related to the initial acquisition of the technology, including:
  - Software costs.
  - Hardware costs, including installation, setup, and the environment (space for equipment, networking, and so forth).
  - Consultation, including in house IT consultation, consultation from vendors, and outside consultants.

- **Maintenance.** This includes ongoing costs, including hardware and software maintenance charges, system administration costs, and upgrades.

- **Training.** This includes initial training for users and system administrators, plus ongoing training as personnel turns over.

For other kinds of projects, you may need to include the cost of hiring staff, manufacturing products, restructuring/converting content, and other incidental expense. Make sure you consider the full range of expenses because you will be grilled on this section by management.

Benefits

This section discusses the financial benefits of the proposal. While it is okay to highlight non-financial benefits, the real purpose of this section is to convince management that the financial benefits justify the cost.

For a cost center business case, you will need to quantify the expected gains. Sometimes this is easy; you can easily quantify how much you will save by shipping a CD instead of a book. Sometimes it is harder; you can estimate how many support calls a particular solution might save, but you will not be able to validate your estimate until after the project has been implemented. Instead, you will need to convince management that your estimate is logical.

For a profit center business case, you will almost always be dealing with estimates. These may be estimates of how well a new product or service will sell or what the price should be. Sales estimates are notoriously inaccurate and pricing exercises are close behind. If you are building a profit center business case, spend time with your company's business people; they will know what management wants to see.

Unless your company is very small, there will be someone who handles financial issues for management. Find that person and spend time with him or her and learn what management wants to see in a business case. Of course, at the same time you should be selling. If you cannot sell the finance person on the value of your business case, you will not have any luck selling management. But, if you can, you will be in a strong position and will have an advocate.

Selling the Business Case

If you write a business case, then just heave it over the transom to your management team, you can bet it will be rejected. Selling your business plan can be just as important as writing it. Here are some suggestions for selling your plan:

- ► **Lay the groundwork:** Well before you submit a business case for anything, you should be building strong relationships with everyone who has a stake in the project. This, of course, includes the financial people mentioned in the last section, but it also includes anyone who will benefit or be harmed by the project.

  This is especially important if you base your argument on cost savings or increased revenue outside your group. For example, if you claim to reduce support calls, work closely with the support organization, both to verify your estimates and to bring them on board as advocates.

- ► **Test the waters:** Circulate your case and test your presentation with people both inside and outside your group. Do not present your case to management until you have worked out the kinks with others. If you have access to a member of the management team that will decide the fate of your proposal, see if you can schedule an informal presentation. Not only will you get useful feedback, you may turn that manager into an advocate.

- ► **Present in person:** You should present your business case to management directly. If you simply send them the written case, your odds go way down. A direct presentation will give you the opportunity to show your enthusiasm and sell, and it will give management the opportunity to ask questions. If the managers give you an immediate okay, great; if not, you can leave them with the written case, then follow up later.

- ► **Use quantitative measures:** Managers, especially those in upper management, will expect quantitative measures. Saying that customers will be happier is fine, but unless you back that up with some kind of quantitative measures, managers will not respond. They want to know that you have done your homework, and to them, homework means dollars and cents.

- ► **Sell to stakeholders:** Sell your proposal to stakeholders beyond the management team. If you propose to reduce support calls by providing user training, make sure the support team wants to reduce support calls, too. If they are swamped with work, they may be happy to get fewer calls, but if not, they may see fewer support calls as jeopardizing their job security. Understand the impact to each stakeholder and factor it into your sales pitch.

- ► **Follow up:** Follow up is crucial. Executives rarely approve or disapprove a business case on the spot. They will want to deliberate. Ask for a date for a decision and then follow up. While you do not want to be a pest, do not let them ignore you either. It is fine to check with them halfway to the decision date to see if they need any additional information or have questions.

Caveats and Limitations

While a business case is an essential tool, it has limitations. Organizations may be run with financial precision and well defined rules and procedures, but their lifeblood is personal relationships. If your management team trusts you, they will accept any reasonable proposal; if they do not trust you, they will reject even excellent proposals.

Ideally, you should have built a strong relationship with your peer organizations and your management team well before you bring a significant proposal to them. Prove that team delivers as advertised and is an integral part of the company's projects. The more you can build strong connections, the easier it will be to build and sell a successful business case.

18

XML Technology

*For a successful technology,
reality must take precedence over public relations,
for Nature cannot be fooled.*
— Richard Feynman

When Feynman said this in the conclusion of his Minority Report Appendix to the Rogers Commission Report [30] on the Space Shuttle Challenger disaster, it was in reaction to what he saw as NASA's unrealistic assessment of the reliability of the Space Shuttle. However, his conclusion applies equally well to any technology, especially those, like XML, that have been heavily touted.

While I am a big supporter of XML technology – I am writing this book using the latest available version of DocBook XML, Version 5.0 – I am also the first to point out that XML is not a panacea. It will not organize your documentation, eliminate your production back-end, or allow you to hire fewer, less skilled, or cheaper writers. But, it does provide the best way to markup nearly all technical documentation.

The Origins of XML

Before I justify that statement, let's take a quick romp through XML's history. XML is the latest in a line of markup languages that originated at IBM in the late 1960s. Charles Goldfarb, Edward Mosher, and Raymond Lorie developed *GML* (originally named using their initials, then later "generalized" to Generalized Markup Language), which became part of several IBM document processing products.

Various kinds of text markup had been common for many years, and GML was just one of many. The critical differentiator for GML was that the developers wanted to, "... restrict markup within the document to identification of the document's structure and other attributes."[16] Goldfarb and his colleagues recognized that if you embed processing commands into your content, you will be forced to update your content when you need to change the processing or if you need to process the content in another environment. Therefore, they replaced explicit processing commands with mnemonic tags that described the content. This allowed them to process the same content with different applications and in different ways without changing the markup.

GML evolved into *SGML*, Standard Generalized Markup Language, which is an international standard, "ISO 8879:1986 Information processing – Text and office systems – Standard Generalized Markup Language (SGML)" [22]. Once again, Charles Goldfarb was closely involved; so closely that he is often referred to as the "father of SGML."

SGML is a "meta-language," that is, a language used to define other languages. Unlike previous markup languages, you do not use SGML directly. Instead, you define a *schema* that describes the grammar. For SGML, schemas are typically defined using a "Document Type Definition" (*DTD*).

In practice, most SGML documents are created using one of the many standard DTDs that have been created for particular purposes. The best known of these is HTML. In the area of technical documentation, the best known and most widely used is DocBook.

SGML was a huge advance from previously available markup languages. By providing independence from any single vendor's processing applications, it spawned a wave of editing and processing applications. And, because several important standards were quickly built on its foundations (HTML and DocBook being among the earliest), it became a great choice whenever interoperability among organizations was needed. SGML also provides the means to create custom grammars for nearly any imaginable subject domain.

Despite its power, SGML in its full glory is complex and can be hard for both humans and software to work with. In the mid-1990s, a working group led by Jon Bosak set out to develop "SGML for the Web," by which they meant a subset of SGML that would be easier to work with and better suited to the Internet. The result was *XML* (eXtensible Markup Language), which is a W3C Recommendation[42]. With some minor exceptions, XML is a proper subset of SGML that sheds the less used and more difficult to implement features of SGML.

Unlike SGML, which was primarily used by documentation specialists for a relatively narrow range of applications, XML hit a sweet spot. It was complex enough to create interesting structures and applications and simple enough that vendors and the open source community could build useful tools and applications to manage data marked up in XML grammars.

XML is used in an amazing array of applications, many of them far from anything imagined when GML was originally conceived. A quick Internet search will yield hundreds of XML initiatives covering topics from aviation to weather. There are more than a dozen music markup languages, even more health care related languages, and too many business information languages to count.

For our purposes, the most interesting XML languages are DocBook, which has moved from being an SGML standard to being an XML standard, and *DITA*, Darwin Information Typing Architecture. Both are standards from the Organization for the Advancement of Structured Information Standards (OASIS).

Key Concepts

Three key concepts lie behind both SGML and XML: schemas, semantic markup, and data independence. [1] I will take a look at these three concepts in turn, using the following example, which is a fragment of DocBook XML taken from the beginning of this section:

---

[1]C. M. Sperberg-McQueen and Lou Burnard's 1994 article, *A Gentle Introduction to SGML* [33] discusses these concepts in detail and provides an excellent starting place for further reading.

Example 18.1. DocBook Example

```xml
<?xml version="1.0" encoding="utf-8"?>
<epigraph xml:id="RFQuote">
  <attribution>
    <personname>
      <firstname>Richard</firstname>
      <surname>Feynman</surname>
    </personname>
  </attribution>
  <para>
    For a successful technology, reality must take
    precedence over public relations, for Nature
    cannot be fooled.
  </para>
</epigraph>
```

Schemas

A schema defines an SGML or XML grammar. There is no single method for documenting schemas. SGML grammars are typically represented using a Document Type Definition (DTD). Some XML grammars continue to be defined using DTDs, but many are now defined through newer schema languages, including RelaxNG and the W3C XML Schema language. While there are technical reasons for choosing one schema language over another in particular circumstances, for our purposes they are all equivalent. I will use the general term *schema* to represent the structure of an SGML or XML grammar, regardless of the particular language used to document the schema.

To see how a schema works, let's start with the first line of Example 18.1. This line, `<?xml version="1.0" encoding="utf-8"?>`, identifies this as XML, version 1.0 of the standard, and defines the character encoding, which in this case is UTF-8, a common encoding for the Unicode character set. You should find this line at the top of every XML file.

The next line, `<epigraph xml:id="RFQuote">`, defines the beginning of an *element* called "epigraph." Elements are the basic structure in XML. Unless it has no content, each XML element has a closing tag. The closing tag for epigraph is: `</epigraph>`. Any content, which can include other elements or text, will be contained between the starting and closing tags.

Elements can be nested, but cannot straddle ("`<a><b></b></a>`" is okay, but "`<a><b></a></b>`" is not).

An XML document is a tree structure, with a single element at the top level surrounding the entire document. For example, in the XML document that is this book, the epigraph is nested inside a `<section>` element, which is nested inside a `<chapter>` element, which is in turn nested inside a `<book>` element. The `<book>` element is the top level element for the document that contains this book. In practice, you can work with sub-trees for editing purposes, and in fact, the various parts of this book are separate XML documents, usually at the chapter or section level, which are combined to create the full book.

In our example, the nesting continues downward from epigraph, which has two elements, `<attribution>` and `<para>` nested inside it. Nested inside `<para>` is content, in this case the Feynman quotation. Nested inside `<attribution>` is `<personname>`, which has nested within it `<surname>` and `<firstname>`, which contain Professor Feynman's name.

The DocBook schema, which this example conforms to, defines each of these elements, plus many more, and the allowable ways in which they can be combined. It also defines a set of *attributes*, which are name/value pairs placed inside element tags to provide additional information about the element. There is an example of an attribute named "xml:id" inside the start tag for epigraph. This attribute assigns a name (in this case "RFQuote") to the epigraph. You can use the name elsewhere in your document to create a cross reference to this location.

As you might guess, there are numerous details – attributes may be optional or required, the order of elements in a particular context may or may not be specified, text may or may not be allowed inside an element, and so forth – but the basic idea is that the schema tells you how to construct the framework of a document and populate it with content. And, it tells applications what they can expect to find when they open an XML document that conforms to a particular schema.

Schemas are the mechanism by which XML allows us to define domain specific languages. This is important because it enables XML to be used for a wide range of applications. Broadening the applicability of this technology has resulted in there being many more tools and applications that process XML than there ever would have been if its applicability had been restricted to a single language.

At the same time, this flexibility means that languages can be devised that meet the specific needs of narrow interests, but still take advantage of the wealth of tools built around XML. Overall, it is a win both for vendors, who get more potential customers, and for users, who get the benefit of standard tools that will work on their customized schemas.

Semantic markup

If you want to emphasize a piece of text in Microsoft Word, you probably just set the font style to italic. Or, in HTML you might use markup like this: `<i>this is important</i>`. In either case, you might think you have identified some text to be emphasized, but in fact you have not. Instead, you have simply identified some text to be rendered in italics.

If you decide later that you want emphasized text to be rendered in some other way, for example, in a bold typeface or in quotations, you need to go back and change the markup for every instance of emphasized text to the new style. But, you cannot do that blindly; what if you also used italics for book titles, or variable names, or any of hundreds of other possible uses? You would need to look at *every* occurrence of italics in the content to be sure you only changed those occurrences where italics were used for emphasis.

Semantic Markup, also called "Descriptive Markup" tags the meaning of various pieces of content without regard to how they might be rendered in various contexts.

In this example, semantic markup for emphasized text might look like this: `<emphasis>this is important</emphasis>`. This markup does not constrain the rendering in any way. If you decide that emphasis should be rendered in a bold typeface or a different font altogether, you can change the rendering without changing the content.

In addition, because the <emphasis> tag is clear about the meaning of the content being marked up, it will not be used to markup other types of content. You can set up your processing engine to render this tag any way you want, and you can change the rendering without touching the content.

Semantic markup is not an inherent trait of XML languages. You can create an XML grammar that specifies rendering in excruciating detail – XSL-FO, which is used to define detailed formatting for print media, is one example. However, for authoring languages like DocBook and DITA, semantic markup is key. Semantic markup makes it possible to target the same content to different output media and enables higher level processing like generating tables of contents, indexes, etc.

Consider Example 18.1 again. The tags here, for example <surname>, clearly identify what the content is, but do not say anything about how that content should be rendered. If you decide that a person's name should be rendered in the Palatino typeface, then decide later that it should be rendered in Arial, you do not need to touch the content.

Processing can go well beyond adjusting fonts. In this example, you could easily choose to render Professor Feynman's name as "Richard Feynman" or "Feynman, Richard." Take it another step further and you could imagine generating a Table of Epigraphs sorted by author for the front matter of a book. If you had simply left the name un-marked, or marked it up with rendering specific markup, you could not have done any of these things.

Data independence

Data independence means that SGML or XML content is independent of the software and hardware used to process it. It also means that data can be shared easily between different environments. Just as semantic markup separates the content of a document from instructions on how to process it, data independence separates the document from the applications that process it.

Data independence comes from XML and SGML being international standards with an open architecture. Vendors can participate in the development of the standard, and they can build applications that use the standard without concern that it will arbitrarily change under their feet.

Many of the most widely used word processing and publishing packages use proprietary data formats. For example, Adobe FrameMaker and Microsoft Word have native formats that, while not completely hidden from view, are proprietary and under the sole control of the companies that own them.

If your content is marked up using XML, it will be easier for you to change your tool set or even use more than one tool set at the same time. It is not at all uncommon for an XML shop to have authors working on the same document using different editing software. As long as the software treats the XML according to the standard, this is easy to do.

XML Pros and Cons

The question for a documentation manager who is not using XML is whether it is worth going to the trouble of moving to XML, especially if your current environment works pretty well. The answer depends on your particular circumstances and is not simply, "if you are doing X, yes, if you are doing Y, no." In this section, I will take a look at the pros and cons of XML.

Reasons to use XML

XML is most useful in situations where content has a lifetime beyond its initial publication, or where content is targeted for more than one deliverable or delivery method. Here is an example where both are true.

I worked for AT&T, and later Novell, in the department that developed documentation for the UNIX Operating System. AT&T licensed the UNIX documentation along with the UNIX software to many different companies. Each licensee took the documentation, modified it, added information unique to their version of UNIX, formatted it to meet their corporate standards, and re-distributed it to their customers.

Originally that documentation was marked up in *troff*, which is a first generation markup language not unlike GML. Troff provides a level of data independence, but it is not a standard format, nor is it particularly descriptive. Further, there was relatively little software available to format output. Therefore, it was difficult for our licensees to customize our content, and it was also difficult for them to take updates from us.

To address these issues, we converted most of our documentation to DocBook SGML, and later XML. This gave our licensees more flexibility in using our documentation. They could, and did, choose different tools for editing and formatting content, and they could much more easily customize their content for their needs.

This is a classic "good" use of XML and illustrates most of the things that XML is good for. Here is a summary of situations where XML works well:

- ► **When you need tool independence.** XML is a widely used and supported international standard. You can choose from a wide variety of tools from many different vendors, some open source, some proprietary. If you use one of the more popular XML languages (for example, DocBook or DITA) you will discover many applications customized to support that language, but even if you use a less popular XML language, or "roll your own," you will find plenty of tools you can work with.

- ► **When you need to reuse content.** If you are designing movie posters, XML is not for you. However, if you are writing documentation that will be updated even a few times over its lifetime, then XML is worth considering.

- ► **When you need to publish the same content in different forms.** XML has a related transform language *XSL* or eXtensible Stylesheet Language. You can create XSL transforms to transform a single XML document into multiple output formats, including print, web pages, help systems, text, and so forth.

- ► **When your content is highly structured.** Most word processing formats will let you define some structure in your content. But, often that structure is illusory, being just names on otherwise flat structures. XML supports both hierarchical structure, through the kinds of nested structures discussed in the section on schemas, and more generalized structures, through linking and inclusion.

- ► **When you need to search your content.** XML allows you to create tags for pretty much anything you can think of. If you choose, you can distinguish a widget from a gadget from a thingamajig, and define five kinds of thingamajigs. This makes it possible for you to search not just on content, but on content in context. For example, you can search for a widget named "Fred" and because you are looking for the name Fred inside a widget, you will not find all the references to Fred Brooks or Fred Flintstone in the bibliography; you will just find the widget named Fred.

▸ **When multiple authors work on the same content.** The XML-based *XInclude* standard allows you to create a document from other documents or fragments of other documents. For example, XInclude will let you select and include a single element and its contents from one document in another document. You can split your deliverables in any way that makes sense to your authors, let them edit using that structure, then assemble your deliverables in any way that makes sense.

Reasons not to use XML

There are not quite as many, but there are a few situations where XML is not such a great idea:

▸ **When your content has a complex layout:** If you are creating documents with complex and precise layout requirements, for example, posters with complex graphics or advertisements, stick with a page layout tool.

▸ **When you create a lot of one-off documents:** If every document you create has unique content and format, so that you cannot reuse either, then XML may not buy you much. In this case, as in the previous case, formatting will probably take as much effort as writing, so again a tool that facilitates formatting will probably give you better mileage.

▸ **When you need to use content from non-writers:** XML is not a great choice when you need to get content from people whose normal job is not writing. The tools currently available require a fair amount of training and are most productive when used frequently. If you try to get engineers or other non-writers to use XML, they will most likely resist. This is changing as tools improve, but for the moment, even if you use XML for your own authoring, you may need to capture outside content in a different format and convert it to XML.

- ► **When your project is very small:** XML is not always the best choice for small projects. As I am writing this, there are are no low-end, out-of-the-box, end-to-end tools for XML. There is no XML equivalent to Microsoft Word.

   However, this limitation is rapidly disappearing. For example, the Oxygen XML editor [http://www.oxygenxml.com/] features WYSIWYG XML editing and a clean interface with print and web formatters, and other products are headed in the same direction. While this is encouraging, unless you are technically savvy and have time to set up the environment, it may be a while before XML becomes your first choice for small projects.

Choosing an XML Schema

If after analyzing your situation and considering the pros and cons you have decided to move to XML, you will need to decide which XML schema to use.   For many organizations, there is not much choice. If you are working with a print publisher that uses the DocBook schema, you will end up using DocBook. If you share a Content Management System (CMS) with another organization, you will probably end up using the same schema they use. And, if you are working in certain industries, for example, aerospace or defense, you may end up using a specialized schema like S1000D.

If you do not have these constraints and are writing documentation for software or hardware products, then your choice will boil down to *DocBook* or *DITA*. These two are by far the most popular and best supported schemas for technical communication. Which one you choose will depend on the following considerations:

- ► What is your content like? Do you write narrative documentation (books, articles, etc.), modular documentation (reference pages, help pages, cookbooks, etc.), or both?

- ► What are your deliverables? Do you deliver printed documentation, web pages, help systems, or all of the above? And, how important is each type relative to the others?

- ► How specialized is your content? For example, do you need to distinguish several different types of product numbers that are unique to your industry, or are you willing to mark up every product number the same way?

- ► How much content do you need to manage and how many writers do you have working on that content?

- ► How much can you afford to spend on tools and support?

Let's take a look at each of these in turn.

Content

The type of content you author is the standard differentiator between DocBook and DITA. Common wisdom is that if you develop narrative material, like books or articles, you should use DocBook, and that if you develop modular documentation like help systems or if you want to build many different deliverables from the same content, you should use DITA.

While this is an important differentiator, you should not use it exclusively. It is possible to write a book using DITA, and it is possible to write modular content using DocBook. Both schemas are flexible enough to handle a wide variety of content.

However, there are important differences in the way the two schemas represent content, which means that while you can represent any kind of content with either schema, you will probably find that one or the other will be a more natural fit for your content.

There are two considerations to think about in finding the best fit:

- ► **Structural markup:** This is the way your writers think about and markup up the overall structure of their content. At one time, most documentation was structured like a book, with front matter, followed by chapters, followed by back matter (appendices, glossary, index, etc). DocBook was initially developed to support this type of structure, though it has evolved over the years to support a much wider variety of structures.

The introduction of help systems integrated with the user interface of a product has led to topic-based methodologies. In a topic-based methodology, writers create individual modules of various types and knit them together into deliverables. (see the section titled "The Tasks" (p. 10) for a high-level discussion of modular methodologies). DITA was designed specifically to facilitate topic-based authoring, and provides explicit support for this mode of development.

- **Inline markup:** This is the detailed markup of content inside your structure. For example, if you are marking up a play, you need to identify dialog and stage directions. If you are marking up a software manual, you need to identify commands, functions, variables, and so forth.

For software and hardware documentation, DocBook and DITA are similar. In keeping with its philosophy, DITA has fewer inline elements and expects users to create what they need as specializations. DocBook has more choices, and is less likely to need specialization. That said, markup for most common software and hardware components is available in both schemas. You may find one or the other more or less to your liking, but the choice will be primarily aesthetic rather than functional.

If you use a topic-based, modular methodology, you will probably find DITA more to your liking. It has native structures that support and "enforce" this style of writing. DITA was developed specifically to support a topic-oriented architecture and its structures steer writers towards that architecture. It is also possible to write a book or an article using DITA, but unless you customize the schema, you will find yourself using elements with names like `<topic>` instead of `<chapter>`.

DocBook does not enforce a particular style, but because of its heritage, it leans towards a more traditional book or article approach to writing. However, it has supported modular methodologies for many years, albeit without the latest terminology. So, it is also possible to write topic-based content using DocBook, but unless you customize the schema, the topics will be marked up using elements like `<article>` or `<section>` rather than `<topic>`.

If you are willing to customize your schema, then you may want to let other considerations have greater weight. If not, then if your team writes traditional documentation, like books and articles, and is not planning to move to topic-based content, you will probably be happiest with DocBook. If on the other hand your team writes topic-based content using a methodology like *Information Mapping*, you will probably be happiest with DITA.

However, many teams need to write both kinds of documentation. If that is your situation, I would lean towards DocBook. I think DocBook does a better job handling modularity than DITA does handling books.

Deliverables

Both DocBook and DITA have standard transforms that let you create a range of deliverables. As I am writing this, both the standard DocBook transforms and the DITA Open Toolkit will produce: print (using XSL-FO), HTML, XHTML, HTML Help (HTML that can be compiled into Microsoft HTML help), *troff*, JavaHelp, and Eclipse help. In addition, DocBook has transforms to generate WordML and plain text, and DITA has a transforms to generate Microsoft's Rich Text Format (RTF). Finally, there are transforms available to convert DocBook to DITA and DITA to DocBook.

Since both DITA and DocBook support essentially the same set of output formats, the important differentiator here is how well the transforms work for your deliverables. The DocBook transforms are more mature, and are well documented in Bob Stayton's *DocBook XSL: The Complete Guide*[37]. The DITA transforms are newer and have less well developed documentation.

No matter which you choose, the odds are overwhelming that you will need to customize the transforms. I have never seen an organization that did not need something different from the standard look and feel. If you have XSL-knowledgeable staff or contractors to customize your transforms, then either DocBook or DITA will serve you well. If you are running on a shoestring, or if you have less experienced staff, you will probably find that DocBook is the better choice. It is better documented and the standard transforms provide a much wider range of parameters and options that do not require any programming.

Customization

Both DocBook and DITA can be used "out of the box" to author useful content, and both can be customized relatively easily.

DITA is designed with customization, or in its terminology "specialization," in mind. It uses an object-oriented model that allows you to create new elements that are specializations of existing elements. For example, if you were writing an application for an airline, you might want to markup flight numbers. Using DITA, the way to do this would be to create a new element called `<flightnum>` that is a specialization of an existing DITA element (for example, the `<prognum>` element).

DocBook starts out with a larger set of elements, so for many applications you can use standard DocBook where you would need to specialize DITA. However, you can customize DocBook in several ways. DocBook supports a user-defined attribute called `role` that can be used to differentiate variant types of an element. The `role` attribute can be used without changing the schema at all, which means that many "customizations," including the flight number example above, can be handled within the standard. If you need to do more, the latest version of DocBook – as of this writing, DocBook 5.0 – provides support for extensive customization.

DITA is distributed as a collection of DTDs, while DocBook is distributed as a RelaxNG schema. While DTDs can be very flexible, RelaxNG provides much more precise control over the schema. This means that even though DITA was designed for customization, and shines in that category, DocBook is also easy to customize.

If you plan to do extensive customization and are working in a modular environment, you will probably be happier with DITA. If you are looking for a schema that will cover a wider variety of content "out of the box," you will probably be happier with DocBook.

Scale

As mentioned in the section titled "Reasons not to use XML" (p. 188), XML may not be your best choice for small, isolated projects. The overhead may simply be greater than any benefit. But, if your project is big enough to consider using XML, but still small, it is likely that DocBook will be more cost-effective.

- ► DocBook is less likely to require customization.

- ► The DocBook transforms are, at this writing, more comprehensive, better documented, easier to modify, and more likely to be useful with minimal modification.

- ► DITA and DocBook can both be used without Content Management software, but as you scale up in size, DITA will need a CMS sooner than DocBook. For a project of any given size, DITA will almost always require more files and more links between files than Doc-Book. Therefore, the point at which managing a project's files manually gets out of hand will come sooner for DITA than for DocBook.

Cost

Except at the low end, it is unlikely that choosing DocBook or DITA will make a significant difference in the cost of moving to XML. The cost of analysing your content, designing and implementing a solution, training your team, and maintaining your solution will overwhelm any likely differential between DocBook and DITA.

There is one potential exception to this. If you need to reuse content extensively, have a large documentation set, and have many writers, DITA *may* enable you to design in efficiencies that would be harder to achieve with DocBook. The only way you will know is by doing detailed analysis of your existing content and your proposed solution. It is not clear to me that DITA provides enough extra support for modular methodologies to make a significant difference over DocBook in this dimension, but I am willing to be proven wrong.

Buzz

There is one other factor you should at least think about; popularity. As I write this, more people use DocBook than DITA, and it is clearly more stable and mature. However, the "buzz" clearly favors DITA. If you go to a technical communication conference, you will find many sessions on DITA and few (or none) on DocBook. There is more and more being written about DITA, and some of the best known consultants are devoting a lot of time to it. Whether this foreshadows a mass movement to DITA and the decline of DocBook is not 100% clear, but there is at least the possibility that DocBook has seen the peak in its popularity.

I would not use this as a deciding factor; in fact, I hesitated to mention it at all. However, if DITA becomes predominant, it will begin to get better tool support and it will become easier to find writers and tools people who are familiar with it. The good news is that if you are using DocBook, there is still a vibrant community that uses it, supports it, and will continue to do so for a long time. And, if you ever need to move from one schema to the other (or to some future schema), it will be much easier to do so with either DocBook or DITA than it would be with proprietary markup.

Putting it together

To make a decision, you will need to weigh the relative importance of each factor for your situation. Usually, the most important factors will be content and deliverables, in that order. Your writers will spend more time with the schema than anyone else; if you pick one that is not a good match in these dimensions, they will be less productive.

The importance of the other factors will vary. If your content would require neither schema to be specialized, that factor will be less important. Or, if you are blessed with an adequate budget, you may not have to stress as much over fine distinctions in cost.

One thing to guard against is letting a secondary factor dominate your decision for the wrong reasons. For example, if you happen to have an engineer who is a DITA expert, that is a plus for DITA, but if everything else points to DocBook, do not let that one factor overwhelm your decision.

A skillful vendor presentation can also have a disproportionate effect. You are best off avoiding sales presentations until you have analyzed your situation and made some basic choices. Then use the presentation as a way to find out how well the vendor's product matches your requirements, rather then letting the vendor steer you towards a solution that matches their product.

In the end, you need to analyze your current situation and your proposed future situation to make sure you understand the core reasons for making a change. Remember that the schema is only one of many decisions you will need to make as you move to an XML-based environment. Look at all of these decisions together, and if you need to, work with an outside consultant to help you better analyze your choices. Be aware, however, that most consultants have their pet solutions, and many have ties to particular vendors. Talk with several before you make a commitment, and look for one who can give you an independent evaluation.

19

Using the Internet

The best Web sites are better than reality.
— Jakob Nielsen

I think Nielsen had his tongue firmly in cheek when he made the statement quoted above, but for tasks and other action-oriented content like troubleshooting information, the Internet can be your best choice. And, you may have no choice; the expense of delivering content in print, which includes not just the cost of printing, but also the cost of packing and shipping, has led many companies to reduce or eliminate printed documentation in favor of web delivery.

This chapter looks at using the Internet for technical communication. I doubt anyone reading this is unfamiliar with the Internet, so I will spare you the tutorial. Instead, I will look at how you can use the Internet and provide some tips for using it more effectively. While I use the terms Internet and web, almost everything described here could also apply to an intranet.

Where are You Starting From?

To get oriented, I will present a progression, from "No-ware" to "Active-ware," that highlights the ways people use the Internet to communicate technical content.

- **No-ware:** No web presence at all. Unless you work for a part of the government that even the NSA does not know about, you are probably not in this category. And these days, even the NSA has a web site [http://nsa.gov].

- **Shovel-ware:** Content is "shoveled" onto the web in whatever form is convenient. This may mean that entire books end up on the web in a single PDF or Word file. While clearly not optimal for all content, Shovel-ware is not always a bad thing. For example, many companies post PDFs of their product manuals that are identical to what was shipped with the product. That way, a customer who has lost the manual can get a printable version on the web.

An example of a good use of Shovel-ware is IKEA [http://ikea.com], which posts PDFs of its assembly instructions on the web.

- **Book-ware:** Content is developed and displayed as books, usually in HTML, with an index and search capabilities. With good search and well designed books, this can be a reasonable way to go, though it does not take full advantage of the power of the web.

Often, Book-ware includes HTML and PDF versions of the same content. This gives readers quick access and easy searching in HTML, plus the ability to print content in a readable form. Hewlett-Packard's documentation web site [http://docs.hp.com] is a good example of Book-ware; they provide parallel HTML and PDF versions of many documents, along with some Shovel-ware for documents like product pamphlets.

- **Block-ware:** Content is designed, developed, and assembled for delivery on the web using a modular methodology. This is the first category that targets content for web delivery and brings the unique considerations of displaying information on the web into play.

Technologies like DITA, and the modular methodologies that most people use with DITA, are good examples of Block-ware.

- **Custom-ware:** Content is designed and developed like Block-ware, but the display of content can be customized by users, either directly by allowing them to select what they want to see or indirectly by having the interface select what they see based on information about the specific product or service being used.

Google News [http://news.google.com] is an example of Custom-ware. Users create a custom web page with news content aggregated from many different sources. A more specialized example is Boeing Aircraft's http://MyBoeingFleet.com service, which gives customers access to customized information for aircraft they own or lease. *Mashups*, which are web sites that combine information from multiple sources into a combined site, can fall into this category if they let users personalize their view.

► **Active-ware:** Wikis, forums, on-line chats, webinars, and so forth. Anything that lets you interact with users falls into this category. This category is orthogonal to the others; you can have interactive features in any environment except No-ware.

There is a continuum within Active-ware, based on the level of in-teractivity, with Wikis at one end, mailing lists and forums a bit further along, twitter [http://twitter.com] next, and webinars/live chats at the other end. There are many good examples of Active-ware, the best known being Wikipedia. In the technical communication world, both DocBook [http://wiki.docbook.org] and DITA [http://dita.xml.org/wiki] have Wikis, and the DocBook community hosts a chat at irc://freenode/docbook.

Other than No-ware, none of these is clearly good or bad. If you simply need to make the assembly manual for a bicycle available for customers to print out, Shovel-ware may be the perfect solution. But, if you have a complex software product that can be configured in many different ways, Shovel-ware may be worse that No-ware.

That said, for most products (even the bicycle), there are some real advantages to the more advanced forms. If you write content for the web and address real user questions, you can significantly reduce support calls and therefore support costs. A well-maintained Wiki or product forum can build a community of committed users who will help each other take advantage of your product or service, reducing support costs and building customer loyalty.

This chapter will look at how you can develop content for the web and how you can exploit the various styles of web content to fit your needs. Let's start by looking how to develop content for the Internet.

Developing Content for the Internet

While you can shovel content onto the Internet, that is far from the best method. People read web pages differently from the way they read most printed material. They tend to scan, look for something that catches their attention, skim that part, and skip the rest.

To create useful web content, you need to write differently than you do for print. I will outline some guidelines for writing for the web, but this is a big topic; for a more detailed exploration, I recommend Jakob Nielsen's work, including: *Writing for the Web*[28] and *Concise, SCANNABLE, and Objective: How to Write for the Web.*[26] Nielsen is down-to-earth and insightful in all of his writings, but his work on writing for the web is especially good. Another good resource is Janice (Ginny) Redish's book, *Letting Go of the Words: Writing Web Content that Works.*[29]

Here are some guidelines that draw on their work and my own experience:

► **Write concisely:** This is good advice anytime, but it is particularly important with web content. In a recent paper[1] Nielsen found that "On the average Web page, users have time to read *at most* 28% of the words during an average visit; 20% is more likely."

Further, Nielsen found that the more words there are on a page, the less time users spend per word. So, as you get more verbose, the percentage of words read goes down.

► **Front-load important information:** Lead with the conclusion, then supply the background. This journalistic technique, also known as the "Inverted Pyramid," originated with the same objective; give readers the most important information first. If readers only spend enough time on your site to read 20% of your content, make sure that first 20% delivers your message.

► **Write strong, short titles:** When readers scan your content, the titles direct their attention and help them select what to read. Good titles lead users to the information that is most relevant to them.

---

[1]Nielsen's paper[27] uses data from a study conducted by Harald Weinreich.[40]

Good titles also help users get to your content via search engines, which give extra weight to words in titles.

Consider the following when you are creating a title:

- **A good title is meaningful:** Titles like "Site Preparation" or "Installation" may work in a book, but on the web they are useless. Instead, try something like "XYZ Product Installation," or "Installing XYZ"

- **A good title leads with the most important information:** The examples in the previous point start with the name of the product or the specific action, which are the most important things in this context.

- **A good title is unambiguous:** I considered titling this book "Leading Writers." The problem is that the word "Leading" can be read as a gerund, which was my intention, or as an adjective, which would imply this was a book about important writers. "Managing Writers" is not as strong a title, nor does it fully carry the implication that the job goes beyond management to leadership; but it *is* unambiguous.

► **Use lists:** Lists organize content for readers, making it easier to understand and remember. That is one reason this book has so many lists. Breaking content into manageable parts does some of the work for the reader.

Where appropriate, title each element of the list. This helps readers scan the content. Plus, good titles make it easier for readers to organize and remember your points.

► **Use task-oriented techniques:** On the web, task-orientation is even more important than in print. Users come to technical documentation to find out how to do something or how to fix something. Clear, concise, and accurate procedures are what they need.

► **Use examples:** Give users detailed examples, plus instructions on how they can tailor those examples for their own use. Many users – and I am one of them – like to cut and paste sample code or command lines; anything that gives them a head start. This is not specific to the Web; it is good practice for any documentation.

► **Use links:** One of the great strengths of the web is the ability to link to other content. Because links are reasonably unobtrusive, there is no good reason to limit your use of them. In technical content, you can link to definitions, prerequisites, reference information, conceptual background, and related topics. In addition, you can link to your feedback page, support lines, forums, related products, and marketing information. If a user might reasonably want to look beyond what is on the page, or if you want to encourage a user to look beyond what is on the page, for example to explore another product, then consider creating a link.

A useful supplement to links is *Mouseover*. A mouseover is some action that takes place when the user leaves the mouse cursor over a link. Mouseovers can be used to display a word or two about the link, pop up a small box with a few lines of content, or pop up a thumbnail of the site you would go to if you clicked on the link. When well-designed, they give users enough information to help them decide if they want to follow the link, and in cases like a pop-up definition, they can avoid the need to click at all.

► **Index your content:** Even with full-text search, indexing is still useful. A good index does much more than just point at instances of a term; it provides context and removes incidental and low-value references.

For example, in this book, the term "XML" appears over 150 times. A full text search would give you all of them, including incidental occurrences like the one in this paragraph. The index in this book points to a handful (less than 10%) of those occurrences, and most of those are further qualified with information like: "Advantages" or "Applications."

► **Optimize for search engines:** Many people will get to your site through search engines like Google, so you will want your site to have a prominent position when users search on relevant terms.

The major search engines rank a web site based on the relevance of its content to the search terms and the "importance" of the web site. The latter is based, usually, on the number of other sites that refer to that web site and the importance of the referring sites. Since these algorithms are complex, proprietary, and constantly changing, it can be difficult to optimize the importance of your site in the eyes of the search engines. However, most search engines give titles greater weight that other text, so use relevant titles, with clear, consistent, and unambiguous terminology.

Getting the Most out of the Internet

This section looks at the more common ways of using the Internet; specifically, Book-ware, Block-ware, and Custom-ware. I will leave Active-ware for the section on Web 2.0. Selecting how far to push your use of the Internet can be tricky; you may be tempted to do more than you really need to do. It is easy to feel like you just *have* to push the limits. But, as with any other technology, take a step back and look at what you really *need,* rather than what you *want.*

Book-ware

Book-ware can be a reasonable compromise between the traditional print-centric approach and the "all in" approach of Block-ware and Custom-ware. Book-ware is distinguished from Shovel-ware by the use of search and indexing capabilities, and in the use of HTML, rather than print-centric formats like PDF or Microsoft Word – though many Book-ware sites include PDF as well as HTML.

To get the most out of Book-ware:

- ► **Keep your table of contents:** The table of contents (TOC) is your reader's point of entry to your content. A well designed TOC groups your content and gives users a high level view of what is there. Full text search is not a substitute for a TOC (or an index).

- ► **Index your content:** This is especially important for Book-ware, where you need to help readers navigate content that may not have been originally developed for the web. Also, make sure your index is easy to find; in a printed book, readers know to turn to the back of the book for an index. That may not be as obvious for an online index.

- ► **Create a master index:** In the early days of the Unix System documentation, there was a separate book that contained a master index to the documentation set. In addition to the obvious advantage to the reader of having one place to start, it also made it easier for authors to re-organize content between releases without being overly concerned that readers would get lost.

Of course, a master index does not remove the responsibility of creating and maintaining a strong overall structure, but it does help readers when your structure is less than perfect or has changed between releases.

▶ **Provide search:** The quickest way to do this is to use Google, which can easily be set up to do customized searches on your content. The best searches, however, require additional setup so that users can narrow their search. For example, if you have content for multiple versions of a product on your web site, users should be able to select which version of the product they want to search.

▶ **Build clear, consistent navigation.** This applies for any web site, but can be particularly critical for navigating around structures that began their existence as books. Make sure it is easy for users to find their product, and their version of their product. As much as possible, keep the structure of content within your product line as consistent as possible, so users do not find installation instructions in an Install Guide for one product and a Configuration Guide in another. The watchwords are clarity and consistency.

▶ **Improve your content.** While you may have chosen Book-ware because you do not have the resources to make a large body of existing content more web-friendly, there are still things you can do to improve your content for the web that do not require a major effort.

This can be an incremental process that starts with the most visible and most frequently viewed parts of your content. Some of the suggestions in the section titled "Developing Content for the Internet" (p. 200) do not require a large investment. To get started, look at your titles and structure. After site navigation, these are going to be the first things a reader encounters on your web site, so they need to be strong. Then, look for places to create links. You may already have textual references to other content. Turn these into links, then look for other places where links make sense.

Block-ware

While it can be a big step for organizations that have large bodies of existing content, Block-ware is the logical starting place for new web-based projects. And, well-designed Block-ware can be used to create usable, if not optimal, print documentation.

To get the most out of Block-ware:

► **Pay attention to structure:** There are three elements to your structure:

- **High level structure:** The overall structure of your deliverables. In the print world, this is the documentation set design – the set of books, their titles, and the high level structure of each book. In the web world, this is the high level structure of your web site.

- **Middle level structure:** The structure of individual modules. A template that defines the structure of a procedure would be one example of middle level structure. In many ways this is the most important set of structures you deal with; if you have consistent structure and markup for modules, you will have more flexibility in combining those modules to create deliverables.

- **Low level structure:** The markup used to create content. In the XML world, this is mostly defined by the XML schema. However, it extends to your use of that schema. All XML schemas contain markup that overlaps in purpose. For example, in DocBook, you can use several elements to represent links. In this book, I chose the `<xref>` element for internal cross-references. There are other elements that I could have chosen, but for consistency, I always use `<xref>`. Maintaining consistent low level structure makes it easier to combine modules and can help maintain – though it does not guarantee – consistency of style.

No matter the level, your structure is critical in Block-ware. It is very easy to dive right into creating modules of content, but without structure, both you and your readers will quickly get lost.

► **Remember that users need context:** This grows out of your structure. Even in the most modularized environments, users need context. In particular, this means that tasks and procedures need to be clear about applicability and prerequisites. In practice, that means being clear about when the task can be used, what version of the product it can be used with, and what needs to be done before and after performing the task.

Creating context can be difficult in modular methodologies because you need to make content work both standalone and in combination with other modules. As you are designing and writing modules, take a step back and look at them from the user's perspective. Ask yourself if the module is self-contained, or if it needs additional explanation? Does it include, or point to, information about how to use the product or service in different contexts? Does it clearly state prerequisites? Context makes the difference between a module that helps the user and one that causes confusion or misleads the reader.

► **Do not over-engineer for reuse.** Content reuse can be a wonderful thing, but it can also be carried to the brink of insanity. Remember that your objective is to increase productivity, not reuse content. Content reuse is simply a tool. Unless you are talking about legal boilerplate, or other content that must be identical wherever it appears, do not try to reuse small bits of content. You will not save much, if anything, in translation (translation memory will catch sentence level reuse), but it will add overhead as writers waste time looking for increasingly smaller pieces of content to reuse.

► **Do not over-do modularity:** Be realistic about how you plan to build your deliverables. It is easy to slice and dice procedures to the point where individual modules are unusable alone. Unless you have a compelling case for reuse, there is no point in creating a module for part of a procedure or part of a reference page. You can always split a module later on if needed.

► **Use, but do not over-use, semantic markup:** Semantic markup describes content based on its meaning rather than its representation. The DocBook `<filename>` element is an example of semantic markup; the HTML `<font>` element is an example of representational markup.

Semantic markup is a critical part of writing for the web. It separates the meaning of a piece of content from the way that content is displayed. Always select a semantic element over a representational tag whenever possible.

While semantic markup is a good thing, it can be carried to extremes. You should only markup differences that matter. Consider the `<filename>` element. You could carry the idea of a file name even further and create elements for the file name of graphic files, commands, devices, databases, *ad infinitum.* In some cases this may be useful – for example if you want to create a list of different kinds of files identified in your content – but in many (probably most) situations, this is overkill and simply adds overhead for no benefit.

Custom-ware

Custom-ware takes Block-ware to the next level by giving the user control, either directly or indirectly, over the display of content. Custom-ware can make it possible for a user to receive content tailored to his or her needs.

To get the most out of Custom-ware:

- ► **Pay attention to user needs:** If you are going to give users control, make sure they can create content based on *their* needs, not yours. Make the choices based on customer visible aspects of the product or service that matter to them.

- ► **Build on a solid Block-ware foundation:** Do not dive into Custom-ware until you have been successful using a Block-ware methodology. While you will be tempted, implementing both at the same time is is asking for trouble.

  Writing good Block-ware lays the groundwork for good Custom-ware. Your writers will be able to see how modules fit together, which structures work and which do not, and they will have a much better idea of what is likely to work for readers.

- ► **Start slowly:** It is easy to confuse users with lots of customization options. Consider starting out with just a few options: maybe product name and version or model number. Then, as you get your feet wet, move on to adding or subtracting optional features. By going slowly you will have time to collect user feedback, both directly and by instrumenting your web site. And, by incrementally adding functionality, you can more easily see how each additional level of customization is received.

- ► **Identify content clearly:** When you give the user the ability to create custom content, you still need to be able to identify the particular content that user is looking at. There is nothing much more frustrating to a support engineer than spending an hour with a user simply trying to figure out what version of the documentation he or she is looking at. A unique identifier placed in a discreet but readable place on every module will make it much easier to confirm what a reader is looking at. Also, make sure your titling gives each module a unique name. Then, with the name and date stamp/version number, a support engineer or your writers can positively identify any content.

- ► **Do not forget privacy:** Storing user preferences between sessions can make your site much friendlier, since users will not need to re-enter information each time they visit your site. However, if you do this, you need to protect their privacy. A discussion of privacy policies is well beyond the scope of this book. Get legal advice and work with your web designer to make sure you properly protect your user's privacy, both in the web site implementation and in your privacy policy.

Custom-ware is at the bleeding edge of technology for technical communicators as I write this, and therefore it can be hard to implement well. I am convinced it will become an important part of the landscape, but it will take time and a lot of mistakes before it becomes mainstream.

Web 2.0 and Beyond

"Web 2.0" is one of those terms that gets batted around by people who want to look knowledgeable about the Internet, even if they know nothing. It is at best ambiguous and at worst misleading. Even the people most closely associated with Web 2.0 are at times maddeningly vague. And, given how fashionable the term is, you can pretty much guarantee that anyone who is doing more than Shovel-ware will claim to have a Web 2.0 solution.

To me, Web 2.0 is nicely encapsulated in the words of Eric Schmidt, Chairman and CEO of Google, "don't fight the Internet." What he is saying is that the power of the Internet and Web 2.0 is in building services that take advantage of the Internet's unique capabilities, rather than simply using it as a portal for the same old stuff.

In practice, Web 2.0 is epitomized by applications like Craigslist, eBay, and Wikipedia, all of which are frameworks that encourage constructive interaction and collaboration among users. In the technical communication world, the closest I have seen to Web 2.0 would be support forums and Wikis; not the cutting edge, but definitely useful.

For technical communicators, Web 2.0 comprises the set of capabilities that characterize Active-ware. That is, capabilities that create a dialog with and among users, rather than just a one-way broadcast of content. Let's look at some examples of Active-ware.

Wikis

A *wiki* is a website that can be authored by users. The best known wiki, Wikipedia [http://wikipedia.org], is an encyclopedia completely authored by users and only loosely managed. There are procedures in place to handle malicious or otherwise inappropriate entries, but for the most part, anyone can create or modify any Wikipedia entry.

For technical communicators, wikis can be used in several ways:

► **Internal communication:** Many organizations use wikis as a means for communicating project information, including plans, schedules, organization charts, contact information, and procedures. You can also use a wiki to host discussions and disseminate news.

Most internal websites are dusty collections of information created because groups think they need to have a web site. A wiki with a clear purpose and active users can breath live into this artifact.

► **Document review:** Putting documents on a wiki for review can be an easy way to gather input. Reviewers can either insert comments or simply change text. Authors can further interact with reviewers by adding an explanation about how they are handling a particular comment. By tracking changes in the wiki, you can gather input in real time, rather than waiting for a reviewer to finish.

▸ **User websites:** wikis are being increasingly used for user websites, especially for open source projects. Both DITA [http://ditawiki.org] and DocBook [http://wiki.docbook.org] have wikis that contain tutorials, FAQs, documentation, and commentary. Typically, user wikis supplement the documentation, though there are cases where the wiki *is* the documentation; Knoppix [http://knoppix.net/wiki], a well-known Linux distribution, is one example.

If you use a wiki for a user website, you need to control access. Some companies limit access to their technical writers and make the wiki read-only to customers. Some have open areas in the wiki, where customers can contribute, along with closed areas. And, some are completely open.

However, even when all or part of a wiki is open, most companies require users to register and login to update content. For example, the DocBook wiki has a list of registered users who have write access. You need to get an administrator to give you permission. Wikipedia requires registration and a login, but no other vetting, as long as you follow their editing policies.

News groups

News groups are a more direct, but less formal means for communication with users. They grew out of email lists, which originally were a one-way broadcast medium. Now, they are more interactive and are often available through both email and websites like Yahoo! Groups or Google Groups. News groups encourage informal, off-the-cuff discussion, but they are not the best place to collect longer-lived information. Their ad hoc structure and the low signal to noise ratio make them unreliable and hard to use as a reference, though they are great for quick responses to unique problems.

An active news group can give you fodder for a Frequently Asked Questions list (FAQ) and insights into the way people use your site. Therefore, if you are going to use news groups, assign some knowledgeable people to monitor them and respond to questions.

Webinars

Webinars are live, on-line courses. Typically, users log onto a website for slides and other visual materials, and either dial up a conference line or connect through the website to listen and interact with other participants. Webinars are often recorded for playback on demand. A well-designed webinar can serve as an on-line training course, too.

While not the best method for delivering detailed procedural or reference information, webinars can be a great way to give users an overview of your product or service and get them started using it. It may also be a good way to publicize new functionality and market the product.

Social networking

Social networking is a more formalized extension of forums and mailing lists. In a well-run social network like The Content Wrangler Community [http://TheContentWrangler.ning.com], you will find a community of interested and informed people who find the interaction with other interested and informed people is worth their time.

Some social networking sites, for example Plaxo [http://plaxo.com] or LinkedIn [http://linkedin.com], are primarily designed to be the business analog of Facebook [http://facebook.com], where people build personal websites to share with their friends. Sites built around a shared interest, like The Content Wrangler, provide additional mechanisms for interaction. Examples include forums, interest groups, articles, surveys, and chat rooms.

Like other Internet capabilities, you can dive very deep into social networking. If you are interested in pursuing this idea, I recommend starting slow, looking at existing sites, and building your community bit by bit.

20

Managing Content

Content is King
— Sumner Redstone

Content Management (CM) comprises everything you do to organize, store, share, and publish content. Content management is critical for the same reason personnel management is critical; you need to manage the things that are important to your business, and the two most important things to you as a documentation manager are your people and your content.

Content management has always been a central focus for documentation managers, but only recently has it gained an acronym and the close attention of software vendors and consultants. Most of that attention has little to do with technical documentation per se. Instead, it comes from the Internet, which has enabled companies to publish nearly anything quickly and easily.

When companies began delivering large amounts of information on the Internet, they quickly recognized the need to manage that information. While this has spawned an industry, that industry primarily wants to sell large, expensive Content Management Systems (CMS) designed to manage diverse content from different sources and publish that content on the web in a consistent, brand-identified format.

On the other hand, content management for technical documentation focuses primarily on managing homogeneous content from fewer sources and publishing that content in diverse media. That media may include the big corporate web site, but it probably also includes help systems and print in various forms.

Both types of CMS focus on managing content, but each emphasizes different aspects of the problem. A corporate, or "Enterprise," CMS will emphasize easy capture of diverse content, web site management, and volume. A CMS for technical documentation will emphasize the development side, including revision control, content reuse, and single-sourcing.

This chapter primarily addresses the second type of CMS. It begins with the concepts behind content management, then discusses content management systems.

Content Management Concepts

Content management boils down to four essentials: organizing, storing, sharing, and publishing your content.

- ► **Structure:** Content structure is the first, and in many ways most important, part of managing your content. Content structure comprises two types of organization: deliverable structure, which is the way content is structured for the customer, and development structure, which is the way content is structured for the writer.

- ► **Storage:** Content storage is a common location where content can be stored and retrieved by anyone on your team. It can be as simple as a set of file folders, or as complex as a high end CMS. The essential elements are a common repository, revision control,[1] and reliable backup.

- ► **Sharing:** Content sharing refers to the procedures that make it possible for more than one person to work on content for the same project. This includes access control, search, linking, markup, workflow, and writing style.

---

[1]This is a controversial point. Some content management systems do not fully support revision control; I consider that a fatal flaw, others do not.

► **Publishing:** Content publishing includes the processes and tools involved in moving content from the development environment into deliverables.

Nearly everything concerning content management fits in one of these categories. The next sections look at them in more detail.

Organizing content

The structure of your content is the cornerstone of your content management strategy. A good structure, especially a good development structure, can help compensate for less than optimal tools, but a bad structure will hobble your efforts no matter how good your tools are.

Like your customers, your writers need to be able to find content quickly and easily, and they need a structure that does not get in their way. This is especially important if you are trying to minimize duplication of content or use a single-sourcing methodology.

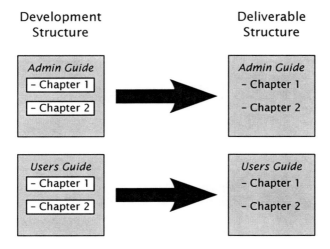

Figure 20.1. Consistent Development and Deliverable Structure

The simplest way to start, and the way that many organizations still work, is to make your development and deliverable structures the same. That is, create a structure for your deliverables and manage your content during development in the same structure. Figure 20.1 shows this kind of structure. The only difference between the development and deliverable structures is that the development structure may be divided into sub-parts (in this case, chapters) for convenience.

Modular methodologies separate the deliverable structure from the development structure. The development structure is typically created around topics, each of which is a separate module. Deliverables are built by combining modules. The way the modules are stored for development is distinct from the way they are combined for delivery to customers.

Figure 20.2 shows an example with distinct structures for development and deliverables. In this case, the development structure consists of topic modules, organized in some way that makes sense for writers. The deliverable structure is build from topics, but organized in a way that makes sense to customers.

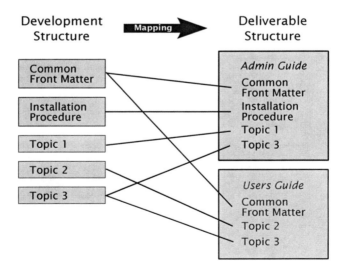

Figure 20.2. Different Development and Deliverable Structure

Separating development and deliverable structures allows you to optimize each for its intended use and also allows you to change one without disturbing the other.

Separation also makes content reuse a little easier. Even if the two structures start out the same, keeping them functionally separate gives you flexibility as your customer needs change and expand. Therefore, I recommend that you create separate deliverable and development structures.

The need to organizing content on web sites has given birth to an engineering sub-discipline, "Information Architecture," a term coined by Richard Saul Wurman.[2] If your content is not clearly structured, or you are unhappy with your current structure, a good Information Architect can help you create both development and deliverable structures. If you are not able to work with an Information Architect, the section titled "Information architecture on a shoe string" (p. 228) provides basic information plus some links to further reference information.

Storing and retrieving content

Once you have a structure for your content, you need a place to put it. This can be as simple as a file structure, (a set of file folders and sub-folders), a file naming convention, and a revision control system. Or, it can be as complex as a high-end CMS. Wherever you are on that continuum, you will need a few indispensable capabilities:

► **A common repository:** You need your content in a place where anyone on your team, including you, can access it.

► **Revision control:** Revision control, also known as Source Control, allows you to store and retrieve successive versions of any file in your repository.

► **Rock-solid backup:** Whether you use a simple file system or a high-end CMS, you need to make sure your content is reliably backed up at least daily.

---

[2] Wurman is probably best known for his Access travel guides (Access London, Access Las Vegas, and so forth) and for his books, *Information Anxiety,* and *Information Anxiety 2.*[44]

Managing content in this way can be a culture shock for writers who are used to keeping all their content on their PC until it is ready to publish. Some writers strongly resist putting partially finished content into a revision control system, possibly out of a fear that it will be pulled out and critiqued by others. But, if you do not keep your common repository up-to-date, it will become worthless.

A good rule of thumb is to never let the gap between the newest work and the contents of your repository get larger than you can afford to slip your schedule. After all, if you lose the content in that gap, it will need to be re-done. In practice, updated content should be checked in at least once a day.

As a corollary to this, make sure it is easy to store and retrieve content. If you are evaluating a CMS, make this a critical requirement. One of the best ways to do this is to use an editor that directly connects with your repository. Most high-end, and a lot of lower-end, editors do this. Regardless of how you implement this functionality, make it fast and easy or writers will "forget" to check-in their work and content will go out of date quickly.

Sharing content

In anything but the smallest organization, writers will need to share content. Sharing content includes content reuse, as discussed in the section titled "Content Reuse" (p. 146), as well as other types of sharing. For example, you may have more than one writer sharing responsibility for a project, or you may simply want to make sure writers can find related content. In any of these cases, you need a mechanism that makes it easy and safe for writers to share content.

Sharing content starts with a common repository for content shared by your writing team, but that is just a first step. Beyond the bare bones, your strategy and systems need to support access control, search, and linking, and your processes need to ensure consistent markup and style.

► **Access control:** Multiple writers must be able to work on the same content concurrently without getting in each others way. Access control can mean locking a piece of content when one writer is working on it or merging changes from more than one writer back into a common source. Either way, the objective is to avoid or manage conflicting edits.

► **Search:** Writers need to be able to find related content easily. They should be able to search the entire repository or browse content following its structure. Writers should also be able to view the status of any piece of content, including who owns the content, what state is it in (for example, draft, final, published), and what products and what versions the content covers.

► **Linking:** Writers need to be able to link both within their own content and with other writers' content. Linking has two dimensions: inclusions, where you pull other content into your own content and links, where you point to other content.

► **Consistent markup:** Writers need to be able to build deliverables that include content from different writers with confidence that the result will be consistently marked up. They also need to know they can build combined deliverables without markup errors.

► **Consistent style:** Writers need to be able to build deliverables that include content from different writers with confidence that the result will hang together stylistically.

While you can manage sharing through manual processes, a good software system will pay big dividends. Even the simplest revision control system will carry you a long way, and a good CMS will make things even easier, especially if you have a lot of content.

Publishing content

Ultimately you need to get your content to users. You need to design a set of deliverables and distribute them. There are two broad categories of publishing, electronic and paper, with a huge array of sub-categories, including help systems of various kinds, web sites, ebooks, printed books, and many more. Depending on what deliverables you need, you may end up transferring content to a traditional publisher in PDF form, a web site in HTML, or a software development team as a help system.

If you use a methodology that separates development structure and deliverable structure, your publishing capability will also need to provide a method for you to specify deliverables by mapping from your development structure to your deliverable structure.

For example, DITA uses a "ditamap," which is file that contains a list of pointers to the modules that make up a particular deliverable. To build a deliverable, the production process finds the correct ditamap, pulls the needed modules from the repository, and builds the deliverable. Some publishing environments will then transmit the completed deliverable to the print publisher, web site, or other partner. In some, typically smaller, situations you may be able to handle everything through a single system, but more often you will need to publish through an external system.

Publishing, especially web publishing, is where you are likely to run into the big corporate CMS. As long as you retain control of the development side, that is probably a good thing; running web sites is not for the faint of heart, nor for those whose primary job is content, not delivery. But, it can be dangerous to think that you can rely on a corporate CMS for development functionality; most of them are not set up to give you good revision control and concurrent access.

I generally think it is better to manage your own content, unless the corporate CMS gives you control over your structure, along with adequate capabilities for storing and sharing your content

Workflow Management and Collaboration

If you have more than one person working on a project, you need some way to coordinate their efforts. Even if you have just one person working on a project, he or she still must work with others, including subject matter experts, reviewers, editors, graphic designers, indexers, and you. While not always considered part of content management, your methodology for managing workflow and collaboration is critical to your ability to manage your content. This section looks into the concept of workflow and means for managing it.

A *workflow* defines how a particular type of work item or deliverable will be completed. It defines the life cycle, from inception through development, test, and delivery. It may also go beyond delivery to include updates, maintenance, and obsolescence. The workflow defines the actions and the actors at each step along the way.

In a modular development environment, a typical workflow for an individual module might include research, design, writing, review (possibly in multiple cycles), revision, test, and delivery. The delivery step could link up with the workflow for a larger deliverable, like a book or website.

Workflow management is the set of processes that define how the workflow will be executed. This includes rules for writers and managers and procedures for ensuring compliance.

A *workflow management system* is computer software that supports workflow management. This typically includes tracking progress, marking status, and generating reports. It may also facilitate the interface with other systems by, for example, sending content from the repository to translators or a publishing system when that content reaches the appropriate point in the workflow.

Every team that has any output at all has a workflow. However, often the workflow comes about by accident. If you do not have a defined workflow, the odds are overwhelming that your workflow is inefficient. If you look closely, you will probably find missteps in your process, for example missing reviews or other quality assurance (QA) glitches. You will also surely find that the workflow varies from writer to writer and between types of deliverables.

Many organizations do not use software to manage their workflow, but whether you use a workflow management system or not, you need to define and manage your workflow. However, you do not need to build a straitjacket; rather, define your workflow in terms of essential steps and let your writers handle the details. This will simplify your tracking and give writers the flexibility to take shortcuts, as long as those short cuts do not eliminate an essential step.

If you decide to use a workflow management system as part of your content management strategy, keep it simple and flexible. For most documentation projects, you only need basic tracking capabilities, like those discussed in Chapter 11, *Tracking* (p. 119). You need to be able to track the status of individual work items and view overall status at any given time.

Workflow Variation

Variation in working styles among writers is common, and not a bad thing. To accommodate different working styles, minimize the number of checkpoints, have clear exit criteria for each checkpoint, and do not worry too much about what happens in between.

If you are managing a large project with many work items and writers, then workflow capabilities, either as part of a CMS or separately, may make sense. However, beware of over complication. I have been managing projects and workflows, both large and small, for many years, and I have seen some pretty complicated processes. Even so, when I looked at the current state of workflow management for this book, I was amazed at the excruciating level of detail and formalization that infects much of the writing about workflow.

Unless you are managing a truly huge project, keep things simple. Build a simple workflow, with only essential steps, manage that workflow with the minimum number of constraints, and if you use an automated system at all, keep things simple. Even with a complex, large project, you will be better off if each separate part of the workflow is as straightforward as possible.

Content Management Systems

Content Management Systems (CMS) have become a large business that comprises a broad range of products and services; everything from simple, inexpensive or free programs designed to support a single web site, to multi-million dollar systems designed to manage millions of pages of content.

CMS Review [http://cmsreview.com] lists hundreds of content management systems, including over 80 open source products, over 130 proprietary products, and over 30 hosted products.

If you listen to the hype, you can easily become convinced that if you are not using the latest high-end XML CMS, you are hopelessly behind the times. But, if you dig a bit deeper, the picture is not quite so rosy; a large percentage of CMS projects fail to meet expectations.

A 2003 report by Jupiter Research[5] found that "more than 60 percent of companies that have deployed Web content management solutions still find themselves manually updating their sites." Assessments by Common Sense Advisory[13] and Computing.co.uk[23] estimate 25 and 40 percent failure rates respectively. Though the numbers vary, there is a message here; adopting a CMS is not easy.

A few themes stand out across these studies:

- ► Organizations consistently underestimate cost and complexity.

- ► Organizations often buy a CMS *before* they define requirements.

- ► Organizations often specify more capabilities than they will ever use, raising cost and complexity unnecessarily.

As you consider whether to use a CMS as part of your content management strategy, start with your needs. There is no question that a CMS can provide most, if not all, of the capabilities you need. However, you probably do not need as many features as you think.

You might not need a CMS at all. Unless you manage a very complex documentation set, you may be able to handle everything you need without a CMS. Complexity, rather than size, is the most important consideration. It is easier to manage a large, but simple, documentation set, than a small, but complex one.

Essential functionality

There is a set of functions that you need to have to successfully manage your content. You do not need to handle those functions with a CMS, but you do need to handle them. These functions include the following:

- ► **Common repository:** You need to have a common place where you and your team can find any content you are working on. The repository can be implemented as a distributed system, with content in different locations, as long as everyone can easily access what they need.

- ► **Reliable backup:** Make sure your backup is rock solid and easy to use. And, make sure your writers update the common repository frequently enough that you will not lose a significant amount of content if you have a failure.

- ► **Revision control:** Some would call this optional for documentation, but I find revision control essential to keep track of previous versions of content. I like to be able to re-create any previous output, and I like the security of knowing I can find anything I have worked on in the past.

- ► **Concurrent access:** Two or more writers need to be able to work on the same content at the same time without stumbling over each others work. A good revision control system will provide this capability and will make it easy for writers to detect possible conflicting changes and merge them.

Additional functionality

In addition to the essentials, the following features are useful, and for some organizations may be essential:

- ► **Workflow management:** Workflow management supports your workflow by keeping track of completion status, schedules, and other information about progress. If you have a large enough project, it can be helpful to have this automated as part of your CMS. Smaller organizations and smaller projects probably do not need this, though they may find it useful.

  Translation management and the publishing interface both need strong workflow management and may justify workflow management software even if the volume is not that large. See the section titled "Workflow Management and Collaboration" (p. 220) for more information.

- ► **Translation management:** If you translate large amounts of content, it can be helpful to integrate translation management into your CMS. Some organizations include translated content in their common repository and give translators access to the repository. If tied into your workflow, this can be an excellent way to connect with your translators. However, you need to be sure they only translate content that is ready for translation (hence the need for workflow management).

- ► **Publishing interface:** An interface with your publishing systems can be a big time saver, especially if integrated with workflow management. For smaller operations you can get away with building deliverables and using a less automated process to publish them, but the ideal for the author is to flag content as complete, push a button, and be finished.

► **Metrics, reports, and alerts:** A CMS can collect all sorts of information, some of it useful, some dangerous. As discussed in Chapter 12, *Measurement and Metrics* (p. 127), I recommend caution in collecting metrics and generating reports. However, your writers can gather useful information like word counts and reuse statistics, which can help them plan their work. And some, sanitized, statistics like reuse, can be useful to you for business cases and presentations.

Of more direct use are alerts. Alerts are triggered by events like reaching a milestone date, changing the status of a content module from draft to complete, or checking in translated content. An alert can trigger email or set a status flag on content. Alerts can be used for normal conditions, for example to tell a translator that a particular piece of content is ready for translation, and for abnormal conditions, for example to tell a writer about a broken link or to flag a missed milestone.

► **Link checking:** A link checker tests links in your content and reports any that are broken. If you do a lot of linking, this capability may become essential.

As with the essential features, you need to consider each of these capabilities. If you do any work at all, you have a workflow and a publishing process, and if you translate, you need to manage that process. But, unless you have a lot of content, the need for automation is less urgent.

Selection guidelines

Selecting a CMS is not much different from selecting any other technology. I would start with Chapter 16, *Acquiring Technology* (p. 157), and supplement it with the following additional guidelines:

► **Beware the hype:** Vendors have a tendency to over-sell "hot" capabilities like single-sourcing and content reuse and under-sell essential, but mundane, capabilities like revision control. Make sure you have the essentials covered before you venture into the heat.

- **Do not build a gold-plated Cadillac:** When I worked at Bell Laboratories, we referred to projects that piled on feature after feature as "gold-plated Cadillacs." As pretty as a gold-plated Cadillac may seem, you do not want a CMS that is burdened with features you will never use. They will just get in the way. Start with the essentials and build from there.

- **Pick a stable vendor:** As I am writing this, the CMS universe is overfilled with vendors while the US economy is headed into recession. Some vendors will disappear and others will merge. This leaves you with two choices: pick a stable vendor with a long track record, or pick an open source technology and bring in consultants or contractors to adapt it to your needs. The one thing you do not want to do is commit to a proprietary technology from a company that is shaky.

- **Use industry standards:** Use industry standards wherever possible. The odds are nearly certain that at some point you will upgrade your CMS, and nearly as certain that you will want to at least consider a different vendor when you upgrade. Vendors, obviously, would rather you stay with them. Therefore, they have less of an incentive to stick with standards, especially if they have a proprietary alternative.

  For XML, this means using a standard schema, but it also means looking out for deviations. There are CMS's that will alter your content in a non-standard way. Some also store metadata in a proprietary structure that can be hard to extract.

- **Drive before you buy:** Get an in-depth demo of any CMS you are looking at, and if at all possible, expand your demo into a pilot project. Even a small pilot project will reveal more than any demonstration.

  Get your team involved in any test drive. Writers will find flaws faster than any manager. It can be useful to have one gungho, fanatic on your evaluation team; he or she will actually read the documentation and get you past minor roadblocks. For the rest of the team, avoid the fanatics and the heel draggers. Instead, look for middle of the road people who are not afraid to express an opinion.

Rolling your own

You may be tempted to roll your own CMS. Nearly every writing group has someone who loves to play with programs and is convinced he or she can pull together a home-grown CMS that will save money and work better than an existing tool.

Resist the urge.

I have managed two groups that had the right skills to do this, but they were the exception, not the rule. And, even though both groups came up with good solutions, I cannot tell you they were more cost effective than a good, off-the-shelf, CMS would have been.

That said, recognize that any CMS project will involve some aspect of rolling your own. You will at a minimum need to deal with the unique characteristics of your environment including, your corporate publication style, your workflow, and your existing content. Even if you contract out everything, you will need to interact with your vendor if you want to get these things right.

So, when I talk about a roll-your-own solution, I am talking about a solution where your team, or an associated IT team, takes primary responsibility for developing or adapting the needed software, rather than a vendor or consultant. Despite all the caveats, there are two situations where you might want to consider a homegrown, or nearly homegrown, solution:

► **Building from an open source solution:** If you have the right skills in house, and you can keep your team from going crazy on customization, you might consider using an open source CMS, maintained by your team. For organizations running on a shoe string, this can be a workable, though potentially perilous, strategy.

  If you do this, apply the same discipline to your internal effort that you would to an external effort. And, plan to spend more time managing the effort than you would spend managing a third party vendor.

► **Essentials only solution:** If you only implement the essentials, and again keep a tight rein on the process, homegrown can work. As with the open-source solution, you should build on proven software, for example SVN (Subversion) for revision control, and keep processes as simple and manual as possible.

If your content is small, you may find that the essentials are all you need, and that you can avoid the stress and strain of implementing a CMS.

Another low cost alternative is a hosted service, that is a service where the CMS is off-site on a computer maintained by a third party and your team accesses it through a web service. If you are not looking for a lot of customization, this can be a cost-effective way to get a CMS.

You can get into a lot of trouble with a roll your own solution; they always take longer and cost more than you planned. However, for a small operation, with a limited amount of content and a simple structure, it can be a viable option, as long as you keep things small and stick to the essentials.

Information architecture on a shoe string

While we are in the clearance sale department of our excursion through content management systems, let's take a quick look at building a development structure on a shoe string.

If your company's business is information or if you have a very large or very complex set of content, then you should invest in your information architecture as a core competency. Otherwise, you may be able to get by with less. This is one place where good enough may really be good enough.

In this section, I will give you a few highlights, but if you need to go it alone, look at Rosenfeld and Morville's *Information Architecture for the World Wide Web*.[31] While this book is targeted at web sites, the basic concepts apply to development or deliverable structure.

Richard Saul Wurman's *Information Anxiety 2*[44] is also useful and entertaining. According to Wurman, there are only five ways to organize information: location, alphabet, time, category, and hierarchy. These conveniently form the acronym LATCH. Here is a quick overview of the five:

- **Location:** Wurman is thinking of physical location here, which might be relevant in describing physical devices. For example, you might organize the documentation for a washing machine based on the location of the controls.

- **Alphabet:** Many topics lend themselves to organization by alphabetical order. For example, you might organize reference material for programming interfaces alphabetically by the interface name.

- **Time:** While more typically thought of as a means for organizing a biography or historical narrative, most procedures are time-based. For example, you typically organize the steps in an installation procedure in the order they should be performed. Organizing content based on the version of a product is also time-based.

- **Category:** This is a broad, and much less well-defined, but essential, dimension. Possible categories for documentation might include type of product, type of task, and type of customer.

- **Hierarchy:** This dimension is characterized by a continuum of some kind. For example, organizing content from beginner to advanced.

In practice, you will probably use several, if not all, of these dimensions to build your structure. For example, you might organize at the highest level based on type of product, then within each category by task, and within each task by time (the order of steps in the task).

However, you do not need to get too fancy. If you are reasonably happy with your current deliverable structure, you may want to start by breaking down your deliverables into modules, looking for commonalities, and building the development structure to incorporate those modules.

The bottom line is that by simply thinking about structure and starting from where you already are, you should be able top build a credible development structure without an expert. You do risk going off into the weeds, but if you are in the much too common situation of having too few resources to do your work, let alone re-design your content, working on a shoe string may be your only option.

21

Avoiding Common Pitfalls

You must learn from the mistakes of others.
You can't possibly live long enough to make them all yourself.

— Sam Levenson

I found out about the decision ten days after it was made. Our organization would be moving to a new content management system by the next February. The decision had been made by a small group selected by and responsible to a higher level manager. For months we had known that this manager, who was new in the position and anxious to make a splash, was dissatisfied with the current tool chain and wanted a "true" Content Management System to improve efficiency and streamline processes. However, we had no idea that there was a search, of sorts, in progress or that a decision had been made.

It should come as no surprise that the resulting system was a disappointment. I left that organization years ago, yet even now, the system is underused and generally despised by those who are forced to use it.

This chapter uses this example to explore some specific pitfalls to avoid when acquiring and using technology.

Misunderstanding or Ignoring Your Real Needs

There is no reason to introduce technology that you have no clear cut need for. Yet, it happens all the time. Many managers cannot resist the lure of whiz-bang technology, whether they need it or not. And salespeople have absolutely no incentive to talk people out of acquiring technology they do not need; quite the opposite.

Just as often, managers may know they need to update their technology, but not really understand their needs. So, they grab the nearest likely looking solution and hope for the best. If you find yourself in this situation, check out the guidelines in Chapter 16, *Acquiring Technology* (p. 157).

My old organization made the second mistake. They needed to upgrade their environment, but they thought that all they needed to do was to drop in place a commercial CMS that was advertised to increase content reuse and enable modular authoring, and everything would be okay. Even though it might have been helpful to increase content reuse and enable modular authoring, the new system could not handle some critical deliverables and did not connect cleanly with our back end processes.

These problems swamped any potential benefit from the new system and left authors wondering why they should bother to use the system at all. While any new system will have growing pains, the perceived benefit to the users needs to exceed the growing pains or they will resist vigorously.

Misunderstanding Your Users

Your users are the people who use or will use the new technology: writers, editors, and managers among others. As discussed in Chapter 7, *Managing Change* (p. 63), your users will resist any change where the perceived benefit *to them* is less than the total perceived pain of adoption.

Understanding your users does not mean kowtowing to their every whim. It means understanding what they really need to do their jobs and making sure that the new technology will make their jobs easier, not harder. It is all too common to presume that users "Don't Know" or "Don't Care," and then abdicate the job of understanding their needs.

This arrogant approach may save you time in the beginning, but it will bite you in the end. Like all humans, the people who use technology do not like to be treated with contempt or condescension. Instead, they need to be treated with respect, brought into discussions of strategy and planning, and made a part of the process.

My old organization took the approach that users did not know enough to be part of the solution; they would only slow down the selection process. This is a big warning sign; whenever you hear that additional input will slow down the process, you can bet that the decision has been made or will be made soon. If this happens after a significant period of requirements gathering, it may be appropriate, even necessary, to close down input and make a decision. But, when it happens early in the process, it is a sign of closed minds.

Misunderstanding Your Requirements

Even if you understand your real needs, you need to translate those needs into requirements. Many projects founder on poorly understood, misinterpreted, or simply non-existent requirements. Take the time to document your requirements, review them, prioritize them, and review them again, before you go any further.

My old organization slipped right by this step with a bit of lip service and a touch of documentation. No one, not even the folks responsible for acquiring the technology, really understood the requirements. As the CMS was deployed and problems were discovered, it turned out that many of the problems could not be correlated to documented requirements.

When you cannot connect a defect in a system with a documented requirement, it is unlikely the vendor will modify the product to your newly discovered requirement as part of the original contract. It is going to cost you extra, assuming they can even meet the new requirement.

This occurs so frequently that vendors write agreements to explicitly cover expanding requirements; usually excluding them without additional consideration. If you are not clear on your requirements or do not carefully document them, it can get expensive fast.

Misunderstanding Your Processes

This one may be a bit surprising, but it is not at all unusual for managers to have only a vague idea of the inner workings of their team's processes. Even if you have well documented processes, there is a good chance that the *real* processes are different. Engineers are always looking for ways to streamline processes to make their lives easier, and their changes do not always make it back into the documented process.

Even when existing processes are well documented and followed, they may not be appropriate in the new environment. However, if you understand your current processes, you will be better able to judge their applicability in the new environment and modify or replace them as necessary. If you do not understand them, you will end up groping around in the dark.

In our example, the biggest process mistake stemmed from an expectation that a wide variety of existing processes – we had well over 100 writers working in four locations on hundreds of documents covering hardware, firmware, and software documentation in print, web, and help output formats – could be replaced with a single, streamlined process. No single person understood all of the existing processes, the new processes were poorly defined, and the transition details were delegated to the users who were expected to figure it all out.

Ignoring Your Intuition

Scott Ambler's *IT Project Success Rates Survey* [1] asked respondents if they had ever been on a project that they knew was going to fail right from the start. Nearly 70% said they had. The bad news is that they could not keep those projects from failing. In fact, when the same group was asked if they considered cancelling a troubled project to be a success, only 41% said yes.

Those two results capture a basic dilemma; we often have at least a gut feeling that things are going wrong, but because we consider cancelling the project to be a failure, we forge on and hope for the best. Likely failure sometime in the future seems to be better than certain failure right now. Of course, that future failure will be more costly and will delay a possible better solution, but it is very hard to take the certain hit now, when there is even a small chance that the project will succeed.

A much better approach would be to acknowledge that gut feeling, then look for its sources. If you can figure out what it is that convinces you the project is headed for failure and can articulate those factors, you have a much better chance of either stopping the project in its tracks or fixing the problems.

Underestimating the Cost of Change

You will underestimate the time needed to adopt new technology, even if you avoid all of the pitfalls mentioned above. Estimating projects that are a known quantity is hard enough; estimating the roll out of new technology is much harder.

In addition to using the techniques described in Chapter 7, *Managing Change* (p. 63), here are a few things you can do to reduce your margin of error. Not all of them apply in every situation, but you should consider each of them:

► **Do a pilot project with a small group:** Not only will this give you an idea of how difficult it will be to deploy the technology more widely, it will give you valuable, early feedback that can be folded into the technology before wider deployment.

► **Phase deployment across your user community:** Extend your pilot to the rest of the user community in phases. By doing this, you can limit the amount of support needed at any given moment and you can once again learn from the experience to make later phases go more smoothly.

Carefully select the users who will be part of each phase. I have seen projects derailed when a group of "Late Adopters" was chosen as the first team to try out a new technology. Let the "Early Adopters" get the first crack. They will be enthusiastic and will help bring along the people in later phases.

► **Phase deployment of features:** Do not try to roll out all of the features at once. Start with the ones that give you the most perceived benefit in return for perceived pain.

► **Keep the lines of communication open:** Especially, do not hide the news from the early phases. The chances are there will be problems; do not try to cover them up. First, you will not be able to; the grapevine is more powerful than any manager. Second, you need to socialize feedback and get ideas for improvements. If you engage the whole team, even those who are not using the technology yet, you will get good ideas, keep everyone enthusiastic, and reduce resistance to change.

22

Epilogue

It is unwise to be too sure of one's own wisdom.
It is healthy to be reminded that the strongest
might weaken and the wisest might err.
— Mahatma Gandhi

Three years ago I left a job as a documentation manager that I had held for several years. I had managed that group in various incarnations over that time, and they had formed a cohesive, experienced team. The company I was working for did not replace me; they left the group in the hands of my manager, at first as an interim measure, then later as a cost savings measure.

At first, I was concerned. If I was a good manager, how could they do without me? I am reminded of a Garrison Keillor monologue where he wished that upon his death there would be a great outcry, women throwing themselves on his coffin, and much wailing and gnashing of teeth. In the same way, I wanted the organization to fall apart and for management to call and beg me to return at an exorbitant salary, or at least to publicly mourn the loss of a great hero.

Of course, none of that happened. The group has done very well over the last few years. They are a largely self-managed team of experienced writers and they have cruised along with minimal problems.

As I have been writing this book, I have come to realize that if things had fallen apart, it would have been my fault. My philosophy, and a central theme of this book has been that managers need to build an environment that empowers their team and leaves it as independent of the manager as possible. If your team requires a minute by minute manager who makes all the decisions and steers through every turn, then you are not doing your job.

As a manager, you should be working yourself out of a job. Not because you want to be laid off or demoted, but because a team that runs itself will be more productive than a team that needs constant management. Plus, a team runs itself will leave you with time to plan your long term strategy, look out for problems, and be a leader, not just a manager.

Becoming an effective leader takes time and experience, and no matter how long the journey, you will never get quite as good at it as you would like. I hope this book will give you some insights to make the journey smoother, but nothing beats getting out there and doing the job. I wish you the best.

Documentation Plan Template

You are free to distribute and use this template, provided you do not charge for distribution or use. An electronic version of this plan in DocBook form is available at the *Managing Writers* companion web site [http://managingwriters.com].

As discussed in Chapter 10, *Project Planning* (p. 101), I have tried to make the written plan as concise as possible. As I have written plans over the years, I keep coming back to the information included here, and rarely find a need for additional sections. If you need other sections, by all means add them, but consider what you really need and avoid fluff.

You can use the template on a per deliverable basis or on a per project basis. Unless the project is very large, I would suggest you use the latter. This will minimize repetition of project-related information and limit the number of documents you need to maintain. If you have a large project, you may also want to consider writing an umbrella document that contains the Executive Overview, Objectives, Overview of Deliverables, Assumptions, and Resources, then using sub-documents for the remaining sections.

Here are some terms used in the template. The term *product* refers to the product, service, or project the documentation is being written for. The term *client* refers to the organization the documentation is being written on behalf of, in other words, the organization paying for your work. The term *user* refers to the end user or customer of the product and its documentation.

Executive Summary

This section summarizes the plan, identifying the high-level deliverables, broad schedule, and level of resources to be used. Its purpose is to reassure the client that you understand what you are going to do, when you are going to do it, and what it is going to cost.

Objectives

This section describes the project and identifies the objectives. It also puts limits on the objectives by defining what will not be covered.

Overview of Deliverables

Describes the type of documentation proposed for this project and identifies the specific deliverables.

Schedule

Describes the schedule from the client's perspective. This should include the following milestones for each deliverable:

- **Design review:** A review of the deliverable design by the client. Some clients will not want to review the design, but I suggest you include it as a milestone and push hard for them to participate. An early review will almost always save trouble later on.

- **Draft reviews:** There will be at least one, and possibly more interim reviews of each deliverable. There may also be separate reviews of different parts. With modular methodologies, there will need to be reviews of each module.

  I recommend scheduling one draft review, with an option for a second review if there are significant problems with the initial draft. If the initial review comments are clear and unambiguous, you should be able to skip a second draft review and go right to the final review.

- ► **Final review:** This is a review of what should be the final version of each deliverable.

- ► **Sign off:** This is the sign off by the client that the deliverable is acceptable for publication.

- ► **Deliverable to production:** The date the deliverable is sent from the documentation team to whoever will produce the final product. There may be multiple dates if you are single sourcing.

- ► **Deliverable published for users:** The date the deliverable is first in the hands of a user (not counting possible user reviews).

- ► **Dependencies:** This includes milestones for deliveries *to* your team or other dependencies. While the previous milestones are roughly in chronological order, dependencies should be interspersed as appropriate.

Depending on your process, you may be able to combine some of the milestones above. For example, if you publish directly to the Internet, the last two dates might be the same. Or, the sign off date might be the same as the date you send the deliverable to production, if that part of the production is automated.

Assumptions

General assumptions that do not have an associated milestone, carry a risk, or need a contingency plan are identified here. These are assumptions like, "We will use XYZ corporation to manufacture printed manuals." Only identify assumptions here if they will be useful to readers or if they identify something that is different from the norm.

Risks and Contingencies

Assumptions that have a milestone, carry a risk, or need a contingency plan are handled here as risks. Risks are documented by stating the assumption at risk, the risk itself, the likelihood of it occurring, the impact if it occurs, and your contingency plan. Each risk is documented using the following format:

Assumption: Statement of the assumption that is at risk.

Risk: What the risk is. For example, "prototype is not ready."

Likelihood: How likely it is that the risk will occur. This can be expressed as "low, medium, or high," or as a percentage.

Impact: The impact of the risk occurring. This should include both an assessment of the magnitude of the impact (low, medium, or high) along with a description of the specific impact. For example, "If the prototype is not ready by date X, documentation development will have to pause."

Contingency: The plan for handling this risk if it occurs. This should include the actions that will be taken and the costs of those actions.

Resources

Describes the resources required to create your deliverables. Depending on what the client wants, this could be expressed in terms of staff months, dollars, or a combination of the two. You may also need to break this down into the kinds of resources required, which may include: a manager, writers, editors, graphic designers, and indexers. You can also identify any special skills or experience required for any position.

Resources might also include expenses for equipment, office space, materials, contract services, consultants, software, IT services, printing, web publishing, and so forth.

Approvals

Here is where you collect signatures or other means of approval from representatives of your client. You may not need to or be able to get every signature on paper in ink, but you do need to obtain and publicly record approvals from all stakeholders and anyone you are depending on.

Glossary

Agile Methodology

According to Wikipedia, "Agile Software Development is a conceptual framework for software development that promotes development iterations, open collaboration, and adaptability throughout the lifecycle of the project."[41]

Agile methods have been around for many years, but the origin of the term "Agile Methodology" dates from 2001, when a group of software developers met at The Lodge at Snowbird in Utah and penned the *Agile Manifesto*.[4]

See also: Extreme Programming, Scrum.

ASCII

The American Standard Code for Information Interchange (ASCII) is a code set commonly used in the English speaking world. It can represent the 26-character alphabet used for most English language words, upper and lower case, plus the most commonly used other characters (for example, punctuation, numbers, and various other symbols).

Because ASCII cannot represent the vast majority of the world's languages – in fact, it cannot fully support English, either, if you consider words like "résumé," which use characters outside the 26-character alphabet – it has been replaced in many contexts, including the XML standard, by Unicode.

See also: Code Set, Unicode.

Attribute

In XML, an attribute is a keyword-value pair inside the start tag of an element that provides additional information about the element. For example, in the following element, the `role` attribute says that this content should be given a `strong` emphasis.

```
<emphasis role="strong">
  some important text
</emphasis>
```

See also: Element.

Audience

The audience is the group of people who will be using your product.

See also: Product.

Code Set

A code set (or coded character set) maps each character of a language to a unique number, which a computer can use in its calculations. For example, the character "A" is represented by the number 65 in the ASCII code set.

The most common code set in use today is Unicode, which maps a large portion of the world's characters into a single code set.

See also: ASCII, Unicode.

Controlled Natural Language

Controlled Natural Languages (CNL) are subsets of the grammar and vocabulary of a natural language designed to reduce ambiguity, improve readability, and facilitate translation. Some CNL's are designed to be completely machine parsable, others are designed to be more easily read by humans. The CNL's of interest to technical communicators are in the latter category and include: Plain English, Simplified English, and Special English.

See also: Simplified English.

Core Competency

A core competency is an area of expertise, skill, or technology that is fundamental to a company or persons activity. Skill at playing the violin is a core competency for Itzhak Perlman. Product design is a core competency of Apple Computer. A core competency typically gives its possessor a competitive advantage; this is certainly the case with the two examples here.

Cost Center

A cost center is a unit within a corporation that is not expected to generate revenue. While a cost center is not expected to generate revenue, it is expected to provide services or other deliverables that are critical to the corporation's success. Within a cost center, new projects are evaluated based on their ability to reduce costs.

See also: profit center.

Critical Dimension

A critical dimension of effort is a part of a job that if neglected will decrease or eliminate customer value. For technical writers, critical dimensions include readability, accuracy, and completeness.

Delegatory Management	Delegatory management is a style of management that delegates nearly all responsibility to the worker. In the work of Robert D. Austin[3] it refers to a management style that delegates measurement and interpretation of metrics to workers.
Deliverables	The tangible things that writers deliver to the project. For example, User Guides, Administrator Guides, Manuals, etc.
Developers	The people who design and build a product. See also: Product.
Development Methodology	A methodology for managing a project, including objectives, schedules, milestones, resources, and so forth. Development methodologies typically fall into two categories: Sequential methodologies define a linear sequence of phases, for example, requirements, design, implementation, test, and deployment. Iterative methodologies define multiple short cycles, typically two to four weeks each.
DITA	Darwin Information Typing Architecture (DITA) is an XML-based architecture for authoring, producing, and delivering technical information. Information can be found at: http://dita.xml.org.
DocBook	A widely used XML grammar designed for developing technical documentation. Information can be found at: http://docbook.org.
DTD	Document Type Definition: used to define an SGML or XML grammar.

See also: Schema.

Element

In XML, an element is the basic structural building block. An element comprises a start tag, some content, which may include other elements, followed by an end tag. The start tag of an element may also contain one or more attributes.

Here is an example of an element named "emphasis" that contains some content and an attribute named "role" with the value "strong."

```
<emphasis role="strong">
  some important text
</emphasis>
```

See also: Attribute.

Environment

The environment is the set of tools, processes, and personnel that a writer works with.

Extreme Programming

Extreme Programming is an agile software development methodology. Like most agile methodologies it emphasizes small releases, open communication, and continuous integration. In addition, it uses the concept of "pair programming," where two programmers work together at one computer, one typing code and the other reviewing the code as it is entered.

See also: Agile Methodology, Scrum.

Flexibility Matrix

A flexibility matrix documents the degree of flexibility a project, or sub-project, has in each of the three planning dimensions (content, time, and resources). It is simply a matrix that orders the three from most to least flexible. While it is a crude measure, it forces the project to consider and document priorities.

Full Supervision

Full supervision is a mode of management where the manager identifies and measures every *critical dimension* of effort from each member of his or her team, and uses those metrics to evaluate employee performance.

GML

Generalized Markup Language. An early markup language developed at IBM by Charles Goldfarb, Edward Mosher, and Raymond Lorie.

Information Mapping

Information Mapping™ is a specialized methodology, developed and owned by Information Mapping, Inc., for analyzing, organizing, and presenting information. Information about this methodology can be found at Information Mapping, Inc. [http://infomap.com]

See also: Specialized Methodology.

Locale

A locale defines a set of user preferences related to location, language, and national conventions. The locale typically defines a language and location (country or territory), plus the character set(s), date and time formats, timezone and daylight savings time conversions, currency formats, and numeric representation.

Mashup

A mashup is a web-based application that takes information from multiple sources and creates a web service that combines the content available from those sources.

Milestone

The term "milestone" originally referred to one of a series of markers along a road, marking each mile along the route. In the project management world, a milestone is an event, either a singular occurrence like "prototype hardware received from Engineering" or the end point of a process like "copy edit completed." In either case, the milestone should be clear and measurable.

Modular Methodology

Modular methodologies for technical communicators decompose content into modules of several different types. Writers develop modules independently, then combine a selected group of modules to create each deliverable. Another common term for modular methodologies is topic-based authoring. Examples include Information Mapping and DITA. While DITA is not itself a methodology, most DITA users follow a topic-based methodology.

See also: DITA, Information Mapping, Topic-based Authoring.

Mouseover

Mouseover is a GUI action that occurs when a user moves the mouse cursor over some position in the interface, but does not click a mouse button. On a web page, a mouseover is commonly used on links and interactive elements like buttons. Mouseover actions on a web page typically display information like the URL for a link, a description of a button's action, related text (for example, a glossary definition), or a thumbnail of the site a link points to.

Partial Supervision

Partial supervision is a mode of management where the manager identifies and measures some, but not all, *critical dimensions* of effort from each member of his or her team, and uses those metrics to evaluate employee performance. Partial supervision leads to dysfunctions in the team by allowing some critical dimensions of effort to be unmeasured and therefore neglected.

Product

The product is whatever you are writing about, even if it is not a product. It could be a service, software, hardware, an airplane, or a toaster.

Profit Center

A profit center is a unit within a corporation that is expected to generate revenue that exceeds expenses. Within a profit center, new projects are evaluated based on expected Return on Investment (ROI).

See also: cost center, ROI.

Repurpose

Content repurposing means that you deliver the same piece of content via different media. For example, if you deliver the same document in print and also on the web, that would be *repurposing*.

See also: Reuse.

Reuse

Content reuse means you put the same piece of content in more than one deliverable on the same output medium. For example, if you maintain a single copy of a glossary definition in source control, then include it in the printed versions of your Installation Guide and User's Guide, that would be *reuse*.

See also: Repurpose.

ROI

Return on Investment (ROI) is the ratio of profit (or loss) relative to the investment for a project. In finance, ROI is typically expressed as a percentage and is projected for several years into the future. The term is often used less formally to describe the expected monetary gain from taking some course of action.

Schedule

The schedule comprises the timeline and milestones for a project.

Schema

A schema defines the grammar of an XML document. There are several languages used to represent schemas, including: Relax NG [http://relaxng.org], the W3C XML Schema Language [http://www.w3.org/TR/xmlschema-0/], and DTDs. These languages each have their strengths and weaknesses in defining any particular grammar. Therefore, standards bodies typically select one of the schema languages to define the normative (official) version of a particular XML grammar. At this time, the normative schema for DocBook is defined using RelaxNG and the normative schema for DITA is defined using a DTD.

See also: DTD.

Scrum

Scrum is an agile software development methodology characterized by multiple short cycles (two to four weeks "sprints"), frequent short communication meetings (called "scrums"), and well defined roles. See http://scrumalliance.org for more information.

See also: Agile Methodology, Extreme Programming.

SGML	Standard Generalized Markup Language. A precursor, and origin, of XML. For further information, see http://www.w3.org/MarkUp/SGML
Simplified English	Simplified English is one of several specialized methodologies that attempt to improve readability, reduce ambiguity, and make translation cheaper and easier. Simplified English defines a set of writing rules and a basic vocabulary that can be supplemented with technical terminology specific to the domain being documented. Related methodologies include: Plain English and Special English. All are examples of Controlled Natural Languages. Information about Specialized English can be found at: http://www.asd-ste100.org/.
	See also: Controlled Natural Language.
Single Sourcing	Single sourcing is a method for reusing or repurposing content to minimize duplication. For example, suppose you have written a procedure for adjusting bicycle chains. If you take the source for that procedure and transform it into a web page and also a printed pamphlet, without altering the original source, you have "single sourced" that content.
	See also: repurpose, reuse.
Specialized Methodology	A specialized methodology is a methodology for some particular aspect of your work. Single-sourcing, Information Mapping, and Controlled Natural Languages are specialized methodologies used by technical communicators.

See also: Controlled Natural Language, Information Mapping.

Tasks

The tasks are the set of things the audience will be doing with the product.

See also: Audience, Product.

Topic-based Authoring

Topic-based authoring is a specialized methodology for content development. A topic is a self-contained piece of content about a particular subject. Topics are authored independently, then combined to create documentation deliverables. Types of topics include conceptual, procedural, and reference. Users of the DITA XML schema typically use a topic-based methodology.

See also: DITA, Information Mapping, Specialized Methodology.

Troff

A document processing system developed in the late 1960s at Bell Laboratories. It was derived from earlier work at Massachusetts Institute of Technology.

Unicode

An encoding system that maps every character in nearly any language to a unique encoding. By providing a unique encoding, Unicode allows text in multiple languages to sit side by side in a document and be processed by any application that understands Unicode. The Unicode Consortium [http://unicode.org] manages the standard, and their web site provides detailed information about the standard and its use.

See also: UTF-8.

Use Case

A description of a user task, usually cast in the form of a person interacting with a system to reach some objective. Use Cases are often used as part of a Requirements Specification.

UTF-8

UTF-8 is a character encoding for Unicode. The Unicode standard identifies characters through an abstract coding that can be implemented in computer systems in many different ways. UTF-8 is the most common character encoding implementation for Unicode.

UTF-8 is the default coding for XML documents, and all XML parsers must support it. It is backwards compatible with ASCII, which makes it easy to use in English only environments. Unless you have unusual needs, UTF-8 is your best choice for character encoding in XML.

See also: Unicode.

Wiki

A wiki is a web site that allows users to edit pages on the site. The best known wiki, Wikipedia [http://wikipedia.org], is an encyclopedia that allows anyone to create or modify entries.

The term Wiki means "quick" in the Hawaiian language. The term WikiWiki, or "very quick" is also used, and is the part of the name of the first Wiki application, "WikiWikiWeb." Additional information can be found at: Wiki (Wikipedia) [http://wikipedia.org/wiki/wiki].

XInclude

The XML-based XInclude standard is a generalized inclusion mechanism that allows you to create a document from other

XML documents or fragments of XML documents. Additional information can be found at: http://www.w3.org/TR/xinclude.

XML

Extensible Markup Language. XML is a specification for defining customized markup languages. Common XML markup languages for technical documentation include: DITA, DocBook, and S1000D. Additional information can be found at: http://www.w3.org/XML.

XSL

Extensible Stylesheet Language. A set of transformation languages that are used to transform XML instances in various ways. XSL can be used to format XML instances into output formats such as HTML or PDF. It can also be used to transform XML instances in other ways, such as generating tables of contents, extracting data, or re-structuring content. Additional information can be found at: http://www.w3.org/Style/XSL.

Bibliography

[1] Scott W. Ambler, August 2007, *IT Project Success Rates Survey:2007*, http://www.ambysoft.com/surveys/success2007.html.

[2] Kurt Ament, 2002, *Single Sourcing:*, Building Modular Documentation, Noyes, ISBN: 0-8155-1491-3.

[3] Robert D. Austin, 1996, *Measuring and Managing Performance in Organizations*, Dorset House, ISBN: 0-932633-36-6.

[4] Kent Beck, et al, 2001, *The Agile Manifesto*, http://agilemanifesto.org.

[5] Matthew Berk, 2003, *Website content management: covering the essentials, avoiding overspending*, Jupiter Research, http://jupiterresearch.com, Summary at: http://www.atnewyork.com/news/article.php/1690881.

[6] Jon Bosak, 2006, *Closing Keynote, XML 2006*, XML 2006 Conference, December 5-1, 2006, Boston, MA, Idealliance, http://2006.xmlconference.org/proceedings/162/presentation.html.

[7] Frederick P. Brooks, 1995, *The Mythical Man-Month*, Essays on Software Engineering, 20th Anniversary Edition, Addison-Wesley, ISBN: 0-201-83595-9.

[8] Marcus Buckingham and Donald O. Clifton, 2001, *Now, Discover Your Strengths*, The Free Press, ISBN: 0-7432-0114-0.

[9] Pip Coburn, 2006, *The Change Function*, Why some technologies take off and others crash and burn, Portfolio, ISBN: 1-59184-132-1.

[10] Alistair Cockburn, 2000, *Writing Effective Use Cases*, Addison-Wesley Professional, ISBN: 0-201-70225-8.

[11] Alistair Cockburn, *Alistair Cockburn's Web Page/Wiki*, http://alistair.cockburn.us.

[12] W. Edwards Deming, 1982, *Out of the Crisis*, Massachusetts Institute of Technology, Center for Advanced Engineering Study, ISBN: 0-911379-01-0.

[13] Donald A. DePalma, 2003, *Rage Against the Content Management Machine*, Common Sense Advisory, http://www.commonsenseadvisory.com/news/pr_view.php?pre_id=4.

[14] R. Stanley Dicks, 2003, *Management Principles and Practices for Technical Communicators*, Longman, ISBN: 0-321-16523-3.

[15] Edsger W. Dijkstra, 1968, *Go To Statement Considered Harmful, Communications of the ACM*, Association for Computing Machinery, Inc., March, 1968, vol. 11, no. 3, 147-148, http://www.cs.utexas.edu/users/EWD/transcriptions/EWD02xx/EWD215.html.

[16] Charles F. Goldfarb, 1996, *The Roots of SGML*, A Personal Recollection, http://www.sgmlsource.com/history/roots.htm.

[17] Anne Gentle, 2009, *Conversation and Community*, The Social Web for Documentation, XML Press, ISBN: 978-0-9822191-1-9.

[18] JoAnn T. Hackos, 1994, *Managing Your Documentation Projects*, John Wiley & Sons, Inc., ISBN: 0-471-59099-1.

[19] JoAnn T. Hackos, 2007, *Information Development: Managing Your Documentation Projects, Portfolio, and People*, John Wiley & Sons, Inc., ISBN: 0-471-77711-0.

[20] Kathy Haramundanis and Larry Rowland, 2007, *Experience Paper – A Content Reuse Documentation Design Experience*, SIGDOC 2007, October 22-24, 2007, El Paso, TX, Association for Computing Machinery, http://sigdoc2007.org .

[21] , 2008, *Rewarding India*, Tradition Meets Transformation, http://www.haygroup.com/Downloads/sg/misc/Rewarding_India_web.pdf.

[22] , 1986, *ISO 8879:1986 Information processing – Text and office systems – Standard Generalized Markup Language (SGML)*, http://www.iso.org.

[23] Neon Kelly, 2007, *High failure rate hits IT projects*, Computing, http://www.computing.co.uk/computing/news/2197021/failed-projects-hit-half-uk.

[24] John P. Kotter, 1996, *Leading Change*, Harvard Business School Press, ISBN: 0-87584-747-1.

[25] John P. Kotter, 1985, *Power and Influence*, Beyond Formal Authority, The Free Press, ISBN: 0-02-918330-8.

[26] Jakob Nielsen, *Concise, SCANNABLE, and Objective: How to Write for the Web*, http://www.useit.com/papers/webwriting/writing.html.

[27] Jakob Nielsen, May 6, 2008, *How Little Do Users Read?*, http://www.useit.com/alertbox/percent-text-read.html.

[28] Jakob Nielsen, *Writing for the Web*, http://www.useit.com/papers/webwriting.

[29] Janice (Ginny) Redish, 2007, *Letting Go of the Words: Writing Web Content that Works*, Morgan Kaufmann, ISBN: 0-12-369486-8.

[30] William P. Rogers, 1986, *Report of the Presidential Commission on the Space Shuttle Challenger Accident*, http://history.nasa.gov/rogersrep/genindex.htm.

[31] Louis Rosenfeld, Peter Morville, 2006, *Information Architecture for the World Wide Web (3rd edition)*, O'Reilly, ISBN: 0-596-52734-9.

[32] , *Scrum Alliance*, http://www.scrumalliance.org.

[33] C. M. Sperberg-McQueen and Lou Burnard, 1994, *A Gentle Introduction to SGML*, http://www.isgmlug.org/sgmlhelp/g-index.htm.

[34] Joel Spolsky, 2004, *Joel on Software*, Apress, ISBN: 1-59059-389-8.

[35] Alexandra L. Bartell, Laura D. Schultz, and Jan H. Spyridakis, *The Effect of Heading Frequency on Comprehension of Print versus Online Information*, Technical Communication, November, 2006, vol. 53, no. 4, 416-426,

[36] William Strunk, Jr. and E. B. White, 1979, *The Elements of Style*, MacMillan Publishing Company, ISBN: 0-02-418220-6.

[37] Bob Stayton, 2007, *DocBook XSL: The Complete Guide*, Sagehill Enterprises [http://sagehill.net], ISBN: 0-9741521-3-7.

[38] Norman Walsh, Leonard Muellner, 1999, *DocBook: The Definitive Guide*, O'Reilly & Associates, Inc., ISBN: 1-56592-580-7.

[39] Norman Walsh, 2009, *DocBook: The Definitive Guide*, O'Reilly & Associates, Inc., Second, ISBN: 978-0596805029.

[40] Harald Weinreich, February 2008, *Not Quite the Average: An Empirical Study of Web Use, ACM Transactions on the Web*, Association for Computing Machinery, Inc., March, 1968, vol. 2, no. 1, article #5, http://doi.acm.org/10.1145/1326561.1326566.

[41] , *Agile Software Development*, http://en.wikipedia.org/wiki/Agile_Project_Management.

[42] , 16 August 2006, *Extensible Markup Language (XML) 1.0 (Fourth Edition)*, http://www.w3.org/TR/xml.

[43] Ray W. Wolverton, 1974, *The Cost of Developing Large-Scale Software, IEEE Transactions on Computers*, IEEE, June 1974, vol. c-23, no. 6, 615-636, http://csdl.computer.org/comp/trans/tc/1974/06/01672595.pdf.

[44] Richard Saul Wurman, 2000, *Information Anxiety 2*, Que, ISBN: 0-7897-2410-3.

Index

Colophon

About the Book

This book was authored using DocBook XML, Version 5.0. It was formatted using a customization of the standard DocBook stylesheets [http://docbook.sourceforge.net], the Saxon XSLT processor [http://saxon.sourceforge.net], and the RenderX XEP Engine. [http://renderx.com].

The XML Press logo and this book's cover were designed by Vicki Fogel Mykles, Write Image, Fort Collins, CO.

You can find more information about this book, including a downloadable version of the documentation plan template, at http://xmlpress.net/managingwriters.html

About the Author

Richard L. Hamilton is principal consultant with R.L. Hamilton & Associates [http://rlhamilton.net], specializing in documentation management and the application of XML technology to documentation. He has managed documentation teams at AT&T, Novell, and Hewlett-Packard.

He has been a member of the DocBook Technical Committee since 2001. He is a contributing author to the DocBook 5.0 Transition Guide and editor of *DocBook: The Definitive Guide, Second Edition*, published by O'Reilly & Associates in cooperation with XML Press.

About XML Press

XML Press [http://xmlpress.net] specializes in publications for technical communicators. We focus on concise, practical publications concerning social media, management, XML technologies, and other topics of interest to technical communicators, their managers, peer engineers, and marketers.

Publications from XML Press

Available now:

Conversation and Community: The Social Web for Documentation
by Anne Gentle

1st Edition, 242 pages with Index,
ISBN: 978-0-9822191-1-9

Conversation and Community brings social media alive through real examples and stories that will help you be a more successful technical communicator. It offers practical guidance and lessons learned from the author's extensive experience as a social media expert and the experience of other industry leaders.

Coming soon:

WIKI: Grow Your Own for Fun and Profit
by Alan J. Porter, scheduled for May 2010.

Best-selling author Alan J. Porter introduces wikis and shows why they are becoming the must-have communications and collaboration technology for businesses of any size. His book includes case studies that highlight the ways that companies use wikis to solve business and communications issues, improve efficiency, and increase customer satisfaction.

DITA Specialization
by Zarella Rendon, scheduled for mid-2010

Zarella Rendon, Senior Consultant and Solution Architect in the Arbortext division of PTC, provides the first in-depth reference for mastering DITA specializations. It describes a detailed approach, with examples and tips from experts along with case studies from successful implementations.

Communicating with Everyone
by Brenda Huettner, scheduled for mid-2010

The US Census Bureau estimates that 27% of the US population has a disability of one kind of another. As a result, responsibility for making product information accessible is quickly becoming a critical part of every technical communicator's job. Industry expert Brenda Huettner addresses accessibility in print, help systems, video, presentations, and on the web. She describes what you need to do to reach everyone in your audience and comply with government regulations in the US, Canada, Great Britain, and the European Union.

http://xmlpress.net

Web site:	http://xmlpress.net
Mailing list:	bookinfo@xmlpress.net
Orders:	orders@xmlpress.net
Twitter:	http://twitter.com/xmlpress (@xmlpress)
Phone:	(970) 231-3624

CPSIA information can be obtained at www.ICGtesting.com
Printed in the USA
BVOW031130011211

277281BV00005B/41/P